ENTREPRENEUR MADNESS

"They all call me crazy, but perhaps I am the only normal one" – Albert Einstein

MARCO ANTONIO SORIANO IV

Published by

BEVERLY HILLS PUBLISHING

468 Camden Drive

Beverly Hills, California 90210

Beverly Hills Publishing Inc.

www.beverlyhillspublishing.com

ISBN: 978-1-7378006-0-6

"To the Ultimate Warrior of Peace, of whom he is but a twinkling reflection. Who has no name yet many, and Who is the source of us all."

—Dan Millman

Acknowledgments

Life has blessed me with many teachers and guiding influences who have, in their own ways, each contributed to the writing of this particular book. Not least among them are my family and close friends who have always nourished me with their love, faith, sacrifice, and belief. I have trusted the inner light and keen instincts from several players in this chess game of life. Thanks to the angels that watch over my shoulders with unsparing support and passionate integrity. You have all created a vibrant launching pad for my first book in our twenty-first century and beyond surely.

From the start, my family and friends provided me an instrumental link between me as the author, the publisher, and the readers. Your excellent support too often goes unheralded, but your efforts were the key that helped me become a published author.

And last, but always first, my abiding love and gratitude to Traccy Jean—my wife, companion, friend, teacher, toughest editor, and most loyal supporter—a blessing in my life and a guardian angel for my spirit.

And, of course, there are my children—Gabriel, Juliette, and Serena—who are my constant aspirations and inspirations to bettering their world.

That is my legacy!

Table of Contents

Chapter 1

Why Do You Think You Are an Entrepreneur? 11

Chapter 2

Is Your Idea Disruptive? 27

Chapter 3

Digitization of Assets: Nouveau Money 39

Chapter 4

Do You Have the Funding? 51

Chapter 5

Putting Together the Idea 63

Chapter 6

Market Research 79

Chapter 7

Branding, Marketing, and Advertising 97

Chapter 8

Sustainability 113

Chapter 9

Delegation: How Many Hats Can You Wear at a Time? 135

Chapter 10

From Entrepreneur to Investor 151

Chapter 11

Understand That People Are Not Numbers 173

Chapter 12

Impact Investment Versus Social Responsibility 189

Introduction

The coming decade is expected to yield many surprises.

- By 2022, only 59 percent of tasks will be completed by humans and 42 percent by machines across twelve industries.[6]
- The 2024 Paris Olympics could see e-sports as an official sport for the first time.[7]
- By 2025, the ICT industry will use one-fifth of the world's electricity, generating more emissions than the entire airline industry.[8]
- By 2025, people will interact with connected devices once every eighteen seconds—that is 4,800 times a day![9]
- By 2025, one in three people on the planet are likely to be overweight or obese.[10]
- By 2027, oil demand and production could reach their peak.[11]
- By 2029, 5G mobile networks could reach their full capacity; 6G may be needed.[12]
- By 2029, AI could be as intelligent as humans.[13]
- By 2030, the rise of global temperatures could breach the 1.5°C target, as humans potentially exhaust the remaining carbon budget.[14]
- By 2030, the number of ICE/diesel cars on the road could plateau.[15]
- By 2030, climate change could push more than one hundred million people in developing countries below the poverty line.[16]
- By 2030, 80 percent of the middle class will be living outside the EU and US.[17]

Sources: [1]UN, [2]WEF 2019, [3]Ovum, [4]Cybersecurity Venture, [5]BofAML Global Investment Strategy, Bloomberg, [6]WEF 2018, [7]BBC, [8]Climate Change News, energuide.be, [9]IDC 2017, [10]World Obesity Fed 2017, WHO 2016, [11]BofAML Global Research, [12]IDC, Ericsson, BofAML Global Research, [13]Ray Kurzweil, as defined by the Turing Test, [14]IPCC 2018, [15]BNEF 2019, [16]Global Commission on Adaptation, [17]Roland Berger.

This book has been on my mind since 2005. Sitting on my balcony overlooking the Long Island Sound in New York at five in the morning, I began to type the first lines. As a professor, I've always had a natural inclination to teach. The old adage, "There are those who do and those who teach," doesn't apply here. I believe that if you're going to teach something—especially what I'm about to teach you—then you ought to have done it first. Agree?

Learning allows you to teach and teaching allows you to learn.

Yes, I've been a professor for most of my career, but that's not what I'm interested in telling you. As a professor, I've worked for various universities around New York. I did it because I loved teaching and because it helped me to master the knowledge applications of my profession.

During my tenure, I've learned that what students taught me was just as important as what I taught them. In fact, they have motivated me to become a better professor, a better teacher. We are all students of life, and we must learn to listen with all of our senses.

That is important because being an entrepreneur means being able to conceptualize your own experiences and translate them into case studies so that others can learn from them. It isn't just about the classroom; it's about the world and your place in it.

I usually begin my lectures with a statement that goes something like this: I hope the competencies I will share with you provide a language for communicating, in concrete terms, about high performance and managerial excellence. I believe that a shared view of the standards we strive to achieve will assist us in our continuing efforts to prepare our ideas, our companies, and our aptitudes for meeting the challenges of the twenty-first century.

Another thing is, as a writer and a public figure, I owe it to myself to pass on my knowledge and experience to others who can benefit from them. In 2005, I wasn't ready. Today, I'm more mature, have more experience, and have many more anecdotes to share. I feel like now is the right time.

This book is a deep dive into entrepreneurship, investing, building relationships, raising capital, growing brands, dominating the market, and most importantly, enjoying what we do for a living. These are all things I've done. You can go to business school, get the textbook lessons on how to do all of these things, and after several years and thousands of dollars, you'll still have to prove that you learned them by putting them into action.

But you'll realize, sadly, that you could have received the same education paying fifty cents in late fees at the local library over the same duration of time. What I'll teach you will go beyond your MBA or doctorate. You will receive an all-encompassing guide to being a successful entrepreneur and investor.

Your legacy. *Your* moves. It starts long before you leave them to the world.

What makes an entrepreneur? There are several good answers to that question and you'll hear them from various quarters. One answer that I like is when someone has worked with their family since an early age. I did that.

An entrepreneur's greatest strength, and a primary key to success, is the quality of staff and skill of the managers. In order to capitalize on this strength, the entrepreneur must create an organizational culture that enables staff to contribute to their maximum potential. Organizational experience shows that, when creating a new culture and building human resources for the future, it is important to define your core competencies. I will discuss this in depth.

A second part of the equation is, is it in you to learn?

Is leading in your DNA? Are you cut out to be independent, and do you have the freedom that an entrepreneur has?

I started working with my family around the age of ten. At one point, my grandfather owned a couple of clinics, a hospital, and a chain of pharmacies, and, by default, a family of medical doctors was all around us. Indeed, I was that little skinny blond boy behind the counter handing out prescriptions and managing the front end of the pharmacy.

My family also owned some restaurants. If I wanted a meal, it wasn't a free meal. I had to work for it. My grandfather would say things like, "You can eat anything you want, but what are you going to do for the restaurant?" If I was going to have lunch, I had to do some work first.

At first, this struck me as weird and sometimes annoying. But as the years went by, I understood to never take anything for granted. I then understood that nothing comes for free.

At the restaurant, I took orders from customers, cashed them out at the register, helped in the kitchen by putting away groceries, and so on. All of that so I could earn a meal. I wasn't pampered.

We used to have these live music events on Friday evenings, and I learned to sing and play instruments with professional musicians. I wasn't the spoiled little kid who got everything he wanted for free. If I wanted it, I worked for it. That enabled me to develop relationship skills and an artistic ear. I loved it.

As I grew older, I began to spend more time with my father, a venture capitalist and renowned community banker. Instead of hanging out in his office, I'd go to meetings and listen. Just listen. People would call him "doctor" even though he never had a medical degree. It was meant as a sign of respect based on the definition of a doctor as "professor."

I'd sit in my father's staff meetings, and other executive gatherings, and learn how to discuss different types of deals, negotiate. I saw the various ways he interacted with lawyers, engineers, architects, politicians, and other professionals, including clients. I was a little sponge, soaking it all in. Eventually, I started doing some of that on my own. Essentially, he put me through an apprenticeship type of training.

Then I got into the higher education system. I briefly recall my high school advisors telling me to apply to several universities, but I said "No." I only wanted to apply to one school—New York University (NYU).

I hated the fact that each application was worth over a grand. I had enough confidence that I would get what I wanted. And I did.

Most people in college don't have a clue what to do. There is an obscure but secretive lack of confidence to say what you want and go for it. Even if it is not the right path. So what? Strategies always change. Remember that!

By that time, I'd already lived experiences most people don't live their whole lives. I had traveled to more than seventy cities on all continents, spoke several languages, and wrote songs and poems.

Early on, I wanted to be a doctor. The motive behind this was the sudden death of one of my uncles, a prominent doctor in our community. He died early in my teenage years. He'd been a strong influence in my life and I felt I needed to pay him respect, honor him. In fact, I was determined.

I was going to be a cardiothoracic surgeon. That is a whole book on its own, what I did during that phase in my life. But suffice to say that while following that path, I realized that I had other aspirations. Stronger ones, where I can make a larger impact on people.

I love mathematics, international business, and people. I love engaging people in communications. College allowed me to expand my way of learning, and I discovered myself. I found literature to be an amazing discipline. Reading books and writing reports on them became a part of my educational experience, and I loved it. In fact, I read most of the major novelists from the eighteenth century, and I read them in romance languages.

It was also an impressive transitional time worldwide as the industrialization era commenced. Multiple political instabilities created a hard identity crisis for millions of people as they were forced to flee their home nations to find safer non-conflictive ones.

Today, we live in similar times. Imagine how that would change your perspective of the world. It made an impression on mine.

Also, during that time, I worked for a few companies not affiliated with my family and realized, believe it or not, that I should not work for other people. It was no surprise to me as I had always been surrounded by entrepreneurial relatives. *I should be working for myself,* I thought.

Possessed with a wild spirit, I couldn't get enough freedom. My biggest flaw, and greatest strength, has always been impetuosity. My friends called me Madness! I guess it is why I titled this book in such a way.

With big corporations, there are hierarchies and protocols to follow. You enter the firm as a junior analyst. Two years in, you're an analyst. Two years later, you're a senior analyst. That wasn't me. It was boring going through that process and waiting for the next step, the next rung on the ladder. Since I'd had a lot of business training at an early age, a lot of it was redundant—the systems and the programs they had in place. Boring. Boring. Boring.

Of course, I'm a creator by nature. I was the kid who walked into a business and started breaking down numbers to understand the costs. I took apart my bicycles and reassembled them to make them faster, smoother, and cooler. Just to understand the ins and outs of things.

Today's children don't develop such curiosity. Technology provides great benefits, but it takes so much more away from creativity, in my opinion.

I wanted to understand the production cycle, and the product, from the inside out. Can it be improved? Can it go global? Can it be localized? Why would anyone want this? Or need this?

I've always analyzed things out on the wing, and I couldn't turn it off. If your mind wanders in this manner, then that is a sign that you may be ready for the next step of entrepreneurship. It is what we call, in academics, *dynamic thinking*. Or, as my pal from the neighborhood would say, "Think outside the box, man!"

Then there's world traveling, which I've done from an early age. Thanks to my entrepreneurial family capable of building global businesses.

It's those experiences that have caused me to have a different way of looking at things. Oh God! If I had the chance to go back in time and do it all over again, I'd do it the same way—except, perhaps, with less partying.

I've always believed that one way to form the world around your own eyes is to learn a foreign language. I speak six so far. All of them are a part of the Roman and Germanic languages. Spanish, Italian, French, Portuguese, English, and German. I'm teaching myself another two, Russian and Chinese. And I'm beginning to study Greek a little more too.

I even speak JavaScript. A little joke for the techies.

I'm blessed to be able to pick up languages quickly. When I visit Mediterranean countries, I can pass myself off as a native because I speak the language properly and understand their culture, and the intricacies of relationships. I can even speak the slang, colloquial sayings that can get you an extra glass of wine. I mention this so perhaps you can see that extracurricular activities can potentially be a great asset to your self-training. As long as you enjoy yourself in the process, you are winning!

That's very important because traveling, experiencing other cultures, and speaking other languages can create multiple opportunities. It opens up some serious doors. And, of course, your perspective changes. Every time you travel, you absorb something of the other culture, something from the location that you visit.

It could be only 1 percent, 2 percent, or 4 percent of the culture. Maybe it's the language that influences you, or the cuisine, or the geography. It enhances your senses, your palate, and your taste for life. Even other people. And your own business.

Today, in my private social clubs, I use this slogan: "Walk in as an American, but come out Italian."

For some, it gives you the impetus to find the love of your life and take your love for a significant other to higher heights and deeper depths.

It's these experiences, and this attitude, that has made me a successful entrepreneur. Now I want to pass that on to you and yours. Similar experiences can be the base on which to build and strengthen human resources in your business, including recruitment, placement, development, and performance appraisals.

Today, with the Internet, we have the ability to learn faster than we ever have before. We can understand the world better. From a consumption point of view and necessity point of view to a desired point of view. These are triggers that indicate how consumers, buyers, and sellers behave. The trends and patterns of consumption allow us to anticipate what you want to buy. Algorithms and artificial intelligence can estimate what you can buy within minutes, if not seconds. It's a bit scary, but it is the reality.

When I get involved in an investment or a new venture, I like to know that the company I'm in with is looking at the different aspects of the business and able to create what I call the culturally accepted product. It involves a mix of the psychological, emotional, and physical, at times even spiritual. Both needs and desires. This is a part of my vision. My investment strategy. I define its competencies, which are a combination of skills, attributes, and behaviors directly related to successful performance on the job.

A culturally accepted product is a product that requires no marketing, no advertising, and no branding because once it has been culturally accepted, it's there for good. People talk about it at parties because it is impregnated in their minds and they can't help it.

Most of the top companies in the world are culturally accepted: Coca-Cola, Apple, and Microsoft, to name a few. If you hear the name or see the symbol, you know what it is, you recognize it immediately. That's powerful. And those are the types of companies I like to get in with or build.

I look for the secret. When is that secret moment that makes a brand culturally accepted? I want to know.

I search for that answer.

Is it surprising to you that all of the emerging markets of the world have one thing in common when sitting down for a meal? Yes, a bottle of soda on the table. Not water. Not wine. Intriguing? If so, then you get it.

Learning. It's the hallmark of an entrepreneur. If you are able to learn from your experiences, and look for experiences that teach you lessons you need to become a better entrepreneur, then you are off to a good start. Learning allows you to teach and teaching allows you to learn. It sounds cliché, but one day you will feel it in your bones to be true.

You can acquire these competencies, but it's not a one-time event. It's an ongoing process. Formal training can help, but experience, coaching, feedback, and individual learning activities are necessary.

Roll up your sleeves. It's time to learn. Ask yourself which of these you have and do not have, then proceed:

- **Core values:** integrity, professionalism, respect for diversity
- **Core competencies:** communication, teamwork, planning and organizing, accountability, creativity, client orientation, commitment to continuous learning, technological awareness
- **Managerial competencies:** leadership, vision, empowering others, building trust, managing performance, judgment/decision-making.

CHAPTER 1

Why Do You Think You Are an Entrepreneur?

You'll meet two types of people in your life. The first person is the one who is comfortable with their job. They're happy in the system, the timing is just right for them, and their lives are on autopilot. We'll call them conformists. They conform to the system that operates around them, and they may not even be aware of the forces acting upon them. They watch shows on Netflix and talk about them in the office.

The conformists are the typical workers. There are a lot of them. In fact, they make up the majority of the people in the world. About 99.99999 percent of the population, I would estimate. I may be missing a few more decimal numbers.

We all have these types in our families or among our friends. They set their alarm clocks and wake up every day at the same time, drink bad coffee to ensure they get to work on time, punch a time clock, take a lunch and two smoke breaks, and head home for dinner. That's the stereotype.

In reality, some of them work second or graveyard shifts, but the routine is essentially the same. They are hardly ever happy with their lives.

The other type of people you'll meet are the pioneers, inventors, makers, creators, angel investors, madness in physical form, or whatever you want to call them. They're the disruptors. Entrepreneurs. They are the .00001 percent. The decimals, as the industry defines them.

But all that jazz aside, ask yourself this question: How do I know if I'm an entrepreneur?

An entrepreneur has a set of characteristics that need to be learned. No one is born with all of these characteristics.

Why do you think you are an entrepreneur?

In practice, these traits relate more to what a person does than what a person knows. They are observable. For each competency, it is possible to describe "behavioral indicators"—actions or behaviors that exemplify the competency in practice. These may be helpful in promoting shared values and common standards of performance and behavior.

Throughout my life, I've met people who talk about what they would like to do, but they never do it. What are they missing?

There is some component, a feature or characteristic, that is missing from their lives that prevents them from doing what they want to do. They punch the time clock when they'd rather invent a new automobile, become a rock star, or plant fruits and vegetables to eat. What's stopping them from moving forward? Is it life choices or something else?

I, for one, say I am an integral person. One who demonstrates the values of my company in daily activities and behaviors. One who acts without consideration of personal gain. One who resists undue political pressure in decision-making. One who does not abuse the power of authority and stands by the decisions that are in the best interest of the business. One who takes prompt action in cases of unprofessional or unethical behavior.

The first step to becoming an entrepreneur is to self-assess. Take inventory of yourself to see which of the characteristics you have and which ones you don't so that you know what you have to learn.

No one is born an entrepreneur. We come into this world with certain characteristics, aptitudes, behavior patterns based on genetics, and so on. But most of that is programmatic. What we're not born with, we pick up from our environment, our parents, friends, families, and so on. Every facet of our lives is defined by these skills and characteristics, our perspectives, desires, and habits that we learn from our environment. We're born and immediately start absorbing information. Some of it good, some of it bad. Some of it is helpful, but some of it is not even worth writing about now.

Let's define professionalism a bit more.

You are professional when you show pride in your work and achievements. You demonstrate competence and mastery of a subject matter, and you're conscientious and efficient in meeting commitments, observing deadlines, and achieving results. You show persistence when faced with difficult problems or challenges. And you remain calm in stressful situations.

Today's children, for example, are tech-savvy at a young age. By the time they're two years old, they can manage a cell phone, tablet, and computer. Thirty years ago, much of that wasn't even available to us. That signals it's also our environment.

We don't know if we would be able to achieve the same level of technological understanding as a three-year-old because we didn't have the opportunity. Maybe today's kids are just smarter, or more dormant. This I do know: the first seven years are extremely important for the toddler to develop their brain neurons and aptitude for learning. So be careful what you are feeding your child's mind.

Another part of that equation, however, is desire. Does the child have the desire to learn? Some don't, and they could be twenty years old and not have a clue about computers. But maybe they're an ace on the guitar. That's what they chose to learn.

Children do learn as they are presented with opportunities, and, soon enough, they show their desires and preferences. Reinforce the strong ones and work easily on the weaknesses. After bringing three children into this world, I think I may be entitled to an opinion on raising them.

An entrepreneur cannot treat all people with a one-shoe-fits-all approach. One must work effectively with people from all backgrounds while treating them with dignity and respect. One should appreciate diverse points of view and demonstrate this understanding in daily work and decision-making.

Picking up a skill doesn't make you an entrepreneur. You can be smart, have a degree, and possess incredible abilities in a certain discipline and still not be entrepreneurial.

Some of my siblings took the safe path. They're low-risk conformists. They have jobs, an education, lovely families, take a mini-vacation every year, and they think they have enough. By contrast, I take four or five months of vacation and can still be effective at doing what I do from anywhere in the world. And, yes, I am their boss when they come to work for me. I'm talking about my older siblings here, by the way.

You have to ask what kind of lifestyle you want. Set that bar and go for it.

A lot of this happens without people realizing it. They wake up one day and they've spent twenty years working as a paper thrower.

It's important to understand what drives you. There are four methods people use to manage their lives. This is what I call the blueprint of the human being. The emotional, the psychological, the physical, and the spiritual. You can have them all or you might just have a few, but it's important to know your makeup.

Your drive comes from those parts of the human soul and to what degree they influence you. It is your makeup, your blueprint. The way you are programmed. Your life. Take charge of it.

For instance, if you're religious, you'll attend church once or twice a week, you may pray a lot, read your Bible or Torah, and perform other spiritual rituals based on your religious background. You may make fun of Tom Cruise and his choice of religion, but you still watch his films. No?

If you're emotional, you might cry a lot at soap operas. Or you're open and transparent, reacting to events on the emotional level.

The psychological level is broad, but it's a strong drive because these people feel a need to prove themselves. We all have these nuts around us. Probably sleeping next to you as you read my book.

The physical is the food you eat and your own health, the exercise you do. Do you eat compulsively or care about the way you look?

These four components also contain differentiations between introverts and extroverts. But you must be aware of them. They may aid you in the self-assessment process.

Everyone has different drives. You have to find out which one is yours. What is your primary driver?

It's also important to revisit the question from time to time because we sometimes forget why we're doing something, what motivated us toward a certain direction. Do you have a strong drive to lose weight or is your motivation to keep your muscles and bones strengthened? Did you miss out on something as a kid? Maybe you grew up poor and now have money so you're overcompensating by indulging in the wrong kinds of foods. Find out what it is that is motivating you and keep that in front of you.

We all need purpose in what we do and the way we do it for ourselves and others.

The reason you want to assess yourself before you take an interest in making money is that it's the natural progression of things. Get the basics down first; then you can advance.

One thing I will emphasize here is accountability. You must take ownership of your responsibilities and honor your commitments. You must deliver outputs and operate in compliance with rules and regulations. You can disrupt the markets as long as you maintain core values.

We've all heard the expression, "Don't judge a book by its cover." That's been said a lot over the centuries, but there's another saying. Judge yourself before you judge someone else. In other words, if you want to be respected, you should respect others.

People in every culture have these expressions, or something similar to proverbs. People intrinsically understand that it's important to look at their own flaws before trying to fix someone else's, and that's why I think the mirror is one of the best darn inventions ever.

Bottom line, figure out where you came from before you decide where you're going.

What Drives You?

In grade school, I took up pencil drawing. In second grade, I'd draw the Ninja Turtles and Marvel Comics characters. My friends were impressed. They'd ask me, "Hey, can I have that?"

"Sure," I replied. To a few of them. For a while at least.

I was drawing those things for me. I didn't want to give them away. But I saw the opportunity, so I started selling the drawings for a few cents here and there.

Then I realized one day that I was making money doing something I enjoyed. I'd buy an extra sandwich at lunch, or a soda. Some days, I'd buy something for a friend. Since then, I've conducted thousands of microtransactions of some kind or another.

Even as a kid, I was constantly scheming, working, negotiating, selling, and being an entrepreneur. But I've moved on from comics to global economies, AI, IoT, EVs, and space travel. Who knew?

Being an entrepreneur is to be a differentiator. An influencer. You want to create a company, a lifestyle, and jobs. It also means helping someone else with their life because it's important to invest in others. Hire talented people and when you influence people, you influence their talent. That's a drive.

Another benefit to being an entrepreneur is you see that your services, your products, and the way you carry yourself touches other people in some way. That's what we call social impact. The more successful you are, the more generous you can be to society. That can be another drive.

I don't care about being famous. I'm not interested in celebrity status. I want to help other people improve their lives and help the environment. That's my theme. You have to find yours.

How do you do that?

To boil it down, it's a thinking process. Ask yourself some questions. Start with these:

- How early do you wake up in the morning?
- Do you shower in the morning or at night?
- When do you brush your teeth?
- Do you exercise?
- Do you want to improve programs or services?
- When do you eat?
- Are you one who offers new and different options to solve problems?
- What do you eat for lunch?
- Do you promote others and their ideas?
- Do you take calculated risks on new and unusual ideas?
- Do you take an interest in others' ideas? Or new ways of doing things?
- What time do you sleep?
- Do you binge? You do; I know you do. But then ask yourself, why? And stop it.

Write it all down. Everything you do for a whole week, write it down. Write down your dreams and your nightmares. Get a journal and at the end of each day, write out what you did that day, when you did it, and how long it took. What did you do from the moment you woke up to the moment you went to bed? This is the first step to understanding your routine, whether you have discipline or not. Use it to understand your standards and the choices you've been making. You see, our minds process thoughts in microseconds.

A great exercise for your brain is to walk from one room to the next and measure the time. OK, it took you ten seconds. Now, close your eyes and do the same distance again in your mind. Apparently, it took you one second to walk that distance. That is how fast our minds are. Sometimes we must learn to slow it down and see things with more clarity.

The point is that you'll learn a lot from that exercise. You might discover that you go to bed at different times every night or that you're not getting enough sleep. You might also discover that when you eat certain foods, you feel better, and some foods make you feel miserable. So you make changes to your life to compensate for your weaknesses.

If you have trouble doing that for yourself, you might have to hire a personal coach to keep you motivated. They'll tell you to run two miles three times a week, do sit-ups, or make a meal plan for you to stay healthy. But you have to start by understanding your choices because choices determine outcomes.

You have to be curious about yourself. That way, you'll see your strengths and weaknesses. Then you can fix them. If you want to run the New York City marathon, you'll need to train. It is all about reaching a balance.

Everything takes training. If you play sports, you need to train. If you don't, you'll come home with a sprained ankle or a pulled muscle. Trainers will tell you those things happen when you don't train and eat properly. It's the same way with being an entrepreneur. There are training regimens that every successful entrepreneur must endure. The first one is self-assessment.

After you've fixed your own issues, you can start helping others. Influencing other people is a byproduct of assessing your own choices and making behavioral adjustments that propel you forward. It starts with being curious about yourself. If you have a certain weakness, find people who complement you. Surround yourself with people who can help you. Hire the talent you need to complement your own strengths and weaknesses so that you can accomplish what you want to accomplish.

Entrepreneurship is about leadership and partnership. It starts with your ability to lead yourself, to be honest with yourself. If you can't do that, then you can't lead others and be honest with them.

The Entrepreneurial Process Is Learning

There's a process to becoming an entrepreneur. It starts with self-assessment. After that, you adjust. But it doesn't end there.

The process is a series of steps you take to accomplish things that matter. It involves your education, business strategy, marketing tactics, relationships you have in the market, and how you assess opportunities. It's easier to understand when you assess your current friendships. How many childhood friends are still your friends? You probably have an inner circle that consists of a best friend and maybe one or two more. Then you have a wider circle that consists of people you work with and have associations with. Then you have another circle of people who are acquaintances, people you're not really close to but you have some connection to.

These circles consist of relationships with people who are necessary for helping you reach your goals. You'll have to ask some of them for favors. Reorganize these circles properly to build your foundation.

Entrepreneurs are notorious for being proud. We think we're the smartest people in the room. Often, an entrepreneur will come up with an idea and think they're the only person to ever have that idea. They think, "Oh, nobody else is thinking like this." That's not the right approach.

It's important to be humble. Asking people for help, for feedback, for constructive judgment. That's how we evolve. We learn by getting feedback from others. That's the process.

In order to do that, you've got to manage your emotions, your actions, your choices. It's all about inner values. It is not who has the best product, but who is first to market.

Pride is internal. You don't learn it. It's intrinsic. It grows from within, starting in childhood. Whether you have a lot of money, you're proud because you have amazing parents, or you delight in your immigrant Italian, Irish, or

African heritage, pride comes from within you but originates from a variety of different levels within your being. Once you have it, you have to translate that into work. You have to use it to influence people. There's constructive pride and destructive pride, so you've got to identify the constructive pride and control the destructive kind.

If you move too fast, take two steps back and reengage. Then you can move forward again. Keep persisting and keep learning. The process involves continuous learning because it's the key to evolution, especially in business.

In the next ten years, half of the companies on the Nasdaq 100 won't be there. These are multibillion-dollar companies, but they'll be dropped from the Nasdaq list because they're not adapting to technology fast enough.

Yes, we saw our favorite toy company go away. Remember it? Every business must have a technological strategy and a safety approach to cybersecurity. It's a part of the process, to learn and adapt. By the way, that was the topic of my first book collaboration in 2019 written by Mathieu Gorge, CEO of VigiTrust, titled, *The Cyber-Elephant In The Boardroom.*

Worth reading it, too.

The Fundamentals of Entrepreneurship

Mark Zuckerberg and Bill Gates attended a top university, then dropped out. Now, both are billionaires. But these are exceptions. Most people who drop out of school aren't geniuses. So do not do that!

Have you ever seen Zuckerberg or Gates in interviews? A bit off-color, are they not?

Education is important. A lot of relationships I have today were nurtured through the educational process. It shouldn't be a luxury. Education should be a part of every society.

If you live in the US and can afford an education, you might attend an Ivy League university. If you don't do that, you might end up in a community college. Either way, what kind of education you receive, from an early age, helps to form your character and social skills.

I don't believe children need to be in school from seven o'clock in the morning until three o'clock in the afternoon, either. Children, if they're having fun, can learn anything. When I was a child, I went to school, got home, ate dinner, did my homework, and went to bed. That instilled discipline and it was good for that, but it wasn't fun. I learned because I was made to learn, but that kind of education doesn't satisfy the soul.

Today, I homeschool all three of my children and they learn because they enjoy the process.

The way we learn today is fun. I like reading and writing books with my kids. We also travel often for months, and that allows them to meet people from all over, which is a learning experience.

Education is necessary because it's where we nurture relationships. It's best when there is a public and private partnership so that people have the ability and opportunity to build relationships that last. Nobody should be denied an education because that's how our character is formed. It allows us to test our ideas and have them criticized.

Another part of learning is teaching. Helping your peers.

As a student, certain subjects were easy for me: physics, chemistry, calculus, mathematics. But these subjects are hard for other people because their brains were wired differently than mine.

I learned to see the opportunity to help others become better students. At one point, in university, I took a job as a student tutor. That allowed me to make money teaching things that I was learning, but it also helped my peers because these were subjects they had trouble understanding. I was able to explain the material in simpler terms than the professor could because I could communicate on a different level.

Communication is vital. As an entrepreneur, you'll have to communicate with different types of people from various backgrounds. Learn to communicate and that will improve your own entrepreneurial skills.

Reading, watching movies, listening to music, all of these activities trigger emotions. There's something psychological that touches you. Some movies make you cry, and some scare you. Others make you reflect.

Experiencing different genres of books, movies, and music can help you learn to communicate better because it becomes a part of the process. It teaches one empathy.

The Competencies of an Entrepreneur

Every entrepreneur has what I call competencies. It's another word for characteristics. I'm not talking about how competent you are at certain tasks. Rather, it has to do with how skillful you are in certain areas of business.

I place them in the following three categories.

Core values: integrity, professionalism, and respect for diversity

Core competencies: communication, teamwork, planning and organizing, accountability, creativity, client orientation, commitment to continuous learning, and technological awareness

Managerial competencies: leadership, vision, empowering others, building trust, managing performance, and judgment/decision-making.

Some people have excellent managerial skills. Others are great communicators. They may be fluent in only one language or multiple languages, but they can communicate. Other entrepreneurs have great influence over others. Some are great delegators. You need these competencies to serve as differentiators. They separate you from the pack. You can have the most brilliant idea, but that doesn't mean you can execute it effectively.

There are times when you must find others with different skills than yours because your weaknesses won't allow you to effectively manage your goals. Maybe you need someone with more maturity, or someone with more experience at something. You need to identify what it is that is holding you back and find the person who has that competency.

Tech entrepreneurs can be cocky. But the process calls for humility. That's why the first step is to learn who you are. Your competencies are going to be different from mine because we have different drives.

If you've ever worked a job and had a boss who was nasty towards you, you might understand. Maybe you quit that job. Well, are you one of those kinds of bosses? Are you patient with people? If not, you might need to work on that because it's a weakness. You'll have more influence with people if you listen to them.

There are a lot of these competencies that define who you are as a professional. You've got to identify what they are for you. What's your blueprint once again?

It's not about making mistakes; no one is perfect. If you make a mistake, learn from that and don't make that mistake again. If you keep making the same mistake over and over again, that's a pattern. Read this book again!

Patterns are hard to break, so it's best to fix it before it becomes a pattern. Toys"R"Us went out of business because they didn't realize that most people were buying toys online instead of in stores. How could a multibillion-dollar corporation not have seen that? Their leaders lacked vision. They didn't understand how the market was changing, and they didn't learn from their mistakes. If you fall into that pattern, you'll lose your business. You've got to adapt.

Learn to Identify Crises and Bubbles

Crises are opportunities. These are times when the market resets. Prices go lower, assumptions are on standby, and you can grow.

The perfect example is the COVID-19 pandemic. People stopped traveling and the hotels took a big hit. Airline companies lost business because people weren't flying while car rentals plummeted. And it trickled down to commercial real estate; it was a domino effect that caused people to lose jobs, companies shut down, and the economy took a downturn. It was a market reset.

A lot of Americans migrated out of states that were politically insane. Yes, you are thinking about it. And I would agree with you, too.

Entrepreneurs, investors see such times as an opportunity.

Don't feel awful about profiting from other people's misfortune. If nobody does anything, the situation will keep getting worse. The investor-entrepreneur can step up and be a hero. You'll invest money, rebuy assets at a cheaper cost, and hire people to help you because you'll still need their skills. Remember, they have competencies you don't have. This is what brings the economy back around and stabilizes it. It is what we call the recovery period.

Crises are a good time to reset the clock. If you have vision, you'll see them as opportunities.

Bubbles are just the opposite. A bubble is created when there is an excessive estimation or expectation of demand. Remember when Elon Musk said his company would accept cryptocurrencies, and forty-five days later, he retracted his words? Well, in doing so, the stock market needle moved quite a bit. Sometimes I wonder about intentions.

A good example of this is what's happening in the electric car sector. Tesla is worth almost a trillion dollars. Now the whole world is trending toward electric cars, so Tesla has created an excessive demand, which has triggered a secondary market. That has led to SPACs, special purpose acquisition companies, allowing people to blindly pool their money into investments they don't understand. People don't even know where the market is going.

Bubbles create an overvaluation of the market. A company that should be worth $100 is worth $100,000. How did that happen? How did this market do so well with no organic growth?

If I make one hundred motorcycles and sell one hundred motorcycles, that makes sense. But if I produce one million cars, and tell you I will sell them all but haven't sold any, that's a bubble.

What I like to do is to go to companies that create these bubbles and tell them they're not worth what they say they're worth. Let's set the record straight, I'll say.

The company is saying they're selling millions of cars and they're not. They're deceiving investors. That's why I decided to build a global electric vehicle platform with my name on it, and with an Italian luxury touch. There's no other electric vehicle company with a living person's name on it, so it's a first.

I build as the demand requests it. It is more expensive but truly practical and unique to the buyer.

We have provenance, history. We can do it the right way.

Organic growth is the best way to show, economically, that a company can be successful. That means I make one hundred bikes today, then I sell one hundred bikes tomorrow. That involves a process. Just because someone knows how to make furniture doesn't mean they know how to sell furniture. You have to find the right customer, the person who actually wants that piece of furniture. You do that by explaining to the customer why that piece of furniture is the golden goose. Otherwise, they can go to IKEA and buy something cheaper and mass-produced.

The key is to create a story, or have a history, that backs up the sales strategy. That's what we're doing with the Soriano brand of multiple products.

Identify the crises, stay away from the bubbles!

CHAPTER 2

Is Your Idea Disruptive?

When entrepreneurs think about inventing, innovating, and solving problems, they don't necessarily think about how they are going to disrupt the market. Instead, they think about how they're going to make something better. It's an improvement mindset.

Disrupting markets means taking shares away from an existing company or group of companies already established in a sector. If your goal as an entrepreneur is to disrupt markets, if that's all you want to do, then you have no means to motivate yourself. That's nothing more than a numbers mindset.

But the entrepreneurial spirit is about creativity. Innovating. Creating. It has nothing to do with numbers or disrupting markets. That could happen, but that's not the way the entrepreneurial mind approaches solving market problems. The genesis of a business isn't there. Think of yourself as an artist.

First and most importantly, the *impresario* (Italian for entrepreneur) wants to be able to show clients and investors that they are adapting. The second thing is, as a service provider and thought leader, the process itself translates into numbers. Investors want to see them and clients want to know how their business is improved. This is embodied in the film *The Founder*, which was about the making of McDonald's and how Ray Kroc took the concept of fast food and turned it upside down.

In the 1950s, eating out was about socializing, hanging out with friends. But Kroc came along and changed the process, the actual event, into simply grabbing a burger with fries and a soda and walking away in five minutes. It was about optimizing the customer's time to ensure people got their food as quickly as possible. It gave birth to a whole new industry—fast food.

Today, McDonald's is one of the top ten companies in the world. One of the first restaurants to open on Mars will probably be McDonald's. His play was simple: own the real estate and make sure the franchise rented it from him for a long period of time.

The Two Types of Entrepreneurs

Entrepreneurial ideas come in two categories. The first, disruptive. The second type of entrepreneur is the complementary one.

Complementary entrepreneurs aren't trying to reinvent the wheel. Whether you're building a car, a motorcycle, a piece of furniture, or whatever, you aren't inventing a new way to do it. The process is already there, the supply lines are set up, and all you have to do is change the style or put your mark on what already exists in the marketplace. You have a tier-one or tier-two supplier you can go to and they cater to you. They craft the parts to your blueprint. If you were creating a new type of wheel, you wouldn't create a square or triangular one. You'd create a round wheel but maybe a different color. You're complementing an industry that already exists.

One advantage to being complementary is it takes less money to start. A disruptive idea requires capital. You need to prove to the market a new concept. Your clients want to see proof of efficiency and the delivery of an idea that works, or a new service that meets a need. Complementary entrepreneurs don't need as much capital because there's a built-in network of clientele.

Take Starbucks. They have different types of coffee on the menu. During holidays, however, they beef up the menu with special flavors. People are attracted to those new flavors, but that's seasonal. It isn't disruptive. It's complementary.

I did have a private conversation once with Jerry Baldwin in an Italian café in downtown Manhattan. I asked, "Why are you here and not at one of yours?" He smiled and, with a Californian accent, replied, "Because you guys still make it better."

I know his inspiration came from his visits to Italy and understanding the importance of coffee in our culture. But he also knows the addiction to sugar and milk derivatives in the United States.

That's one way to complement a market, by introducing seasonal products. That happens a lot in the food industry. A seasonal menu item allows customers to try something and then look forward to it the next year.

When you're inserting yourself into an existing network of supply lines, there's no need to reinvent the wheel. When creating a new motorcycle or scooter, the first thing you have to figure out is how many parts you're going to need, what kind of parts you'll need, and what materials you'll want those parts to be made of. You may not manufacture all of the parts yourself. But once you define the basics, you can set up your network of suppliers, engineer the product, and send your suppliers the design specifics they need to produce the parts you'll require in the desired time frame.

By not reinventing the wheel, you align your company to work with established suppliers before you go to retail. The supply chain, and the economics of the supply chain, haven't changed. The economy of scale already exists and predetermines the way you align yourself to the market. That minimizes your costs.

If you don't have to hire one hundred employees to make bolts and buttons for your product, then that's capital you don't need. You can put that money into suppliers that already have processes and resources in place.

One thing I do differently is that I invest in the suppliers. That way, they can see that my commitment is in their company and I'm not just using them as a service provider. I actually love to own equity in the supply lines that I'm using. In essence, it's a spiderweb structure of a global corporation. Not a hierarchy. And this creates the perfect exit strategy for any investor. I will discuss this in more detail in another chapter.

Entrepreneurs, Risk, and Techceleration

Entrepreneurs are risk-takers. In fact, the ability to take risks is the underlying foundation for an entrepreneur, and an investor. In Western cultures, risk is defined as something completely negative: a noun, a situation involving exposure to danger. I'm already scared. But in Eastern cultures, risk is a balance of life. No wonder the one hemisphere with two-thirds of the world population is doing better economically.

High-net-worth individuals are people with more than $1 million in liquid assets, not including the value of their home. The middle class falls below that. Economists and other business thinkers like to discuss the difference between Wall Street and Main Street. The middle class, as of 2021, makes up more than 50 percent of the world population. How do they differ? Why do they exist? Can you comprehend the impact of these stats?

The people with the most leverage are the high-net-worth individuals. They're able to take on opportunities that everyone else can't. They can take more risk and take bigger risks than the middle class. They expect to get a bigger return based on the amount of risk they take. And the wealthy understand this.

Those of the middle class are mostly laborers. These are the Main Street investors. They tend not to take on as much risk. They play it safe. Conformists. And they're usually more conservative. The reason they seldom get rich is that they make safer investments. They can go their entire life and end up with a small retirement fund. I watched a show where comedian Chris Rock said, "I am rich, but the guy you don't know that pays me to be here is wealthy."

How do you break that barrier? The answer is with risk. You can take some risk, see what the market is evolving into, and adapt to that. That's what I call techceleration. (Yes, I coined it. Use it at your next cocktail party and quote me.)

Techceleration is the rapid advance of technologies available to us today. Unless you live in an emerging market, access to computers, Wi-Fi, and the Internet is common. In the US, more than 90 percent of us are connected to the Internet. We have computers, tablets, smartphones, and other devices that keep us connected to the Internet. There is no reason we shouldn't adapt these to our personal lives.

Businesses connect so they can move faster, provide faster and more efficient services, retain faster payments, and so on. That's why the fintech sector is huge. And some of the ways cryptocurrencies are used are amazing. Robotics are incredible.

Risk will always be the definitive factor between high-net-worth individuals and the middle class. That's true of investors and entrepreneurs.

A part of that is innate. A part of it is personality. Some people are naturally aggressive, social, extroverted. They are the risk-takers. The opposite of that is the conservative, introverted, and shy people. They're a part of our families, business colleagues, and other acquaintances.

The other part of it is learning. You've probably encouraged a friend to take a chance or to give something a try. Maybe someone has encouraged you.

It's easy to identify these characteristics in people. When you eat, for instance, there are certain foods you are not attracted to because they may not look appetizing, or they might smell funny. That stops you in the moment and you may not be ready to take the risk, to try a new food out. You might have to teach your palate to like the taste. Some people criticize foods they aren't familiar with. They talk poorly about certain types of consumption or lifestyles before they've tried them. There is something in every human being, in our minds, that causes us to either take a chance on something new or criticize something before we try it. If you're one of those people who criticize before you try something, it's time to flip the script. Get on board with taking risks. That's the only way you can expect higher rewards.

Obviously, that's not for everyone. If you're content with what you've got, where you are in life, that's fine. Keep it that way. You won't develop a tolerance for risk that way, but some people are wired in that manner and don't want to change. This book is designed for you to unlearn old habits and restart the engine.

Risk can be learned and balanced. If you play a sport, you get better with continued training and practice. Some people are innately talented in some way, but they still need to train. It's essential to begin with identifying the traits that make up who you are.

Take, for instance, two of the greatest football players today. Lionel Messi and Cristiano Rolando. Both are extremely talented, but one trains to death and the other barely runs on a pitch. Which one is more successful in your mind?

Start with defining who you are because running a business means eating it, digesting it, and letting it fuel you. Learning is a process of its own that involves risk and leads to more risk-taking.

So which industries are going to be disruptive in the next decade? I'll discuss those now.

Disruptive Industry Sectors

These are the disruptive industry sectors worldwide for the next ten years.

The paradigms of the 2020s have already started. These are the ones I've identified for the next decade.

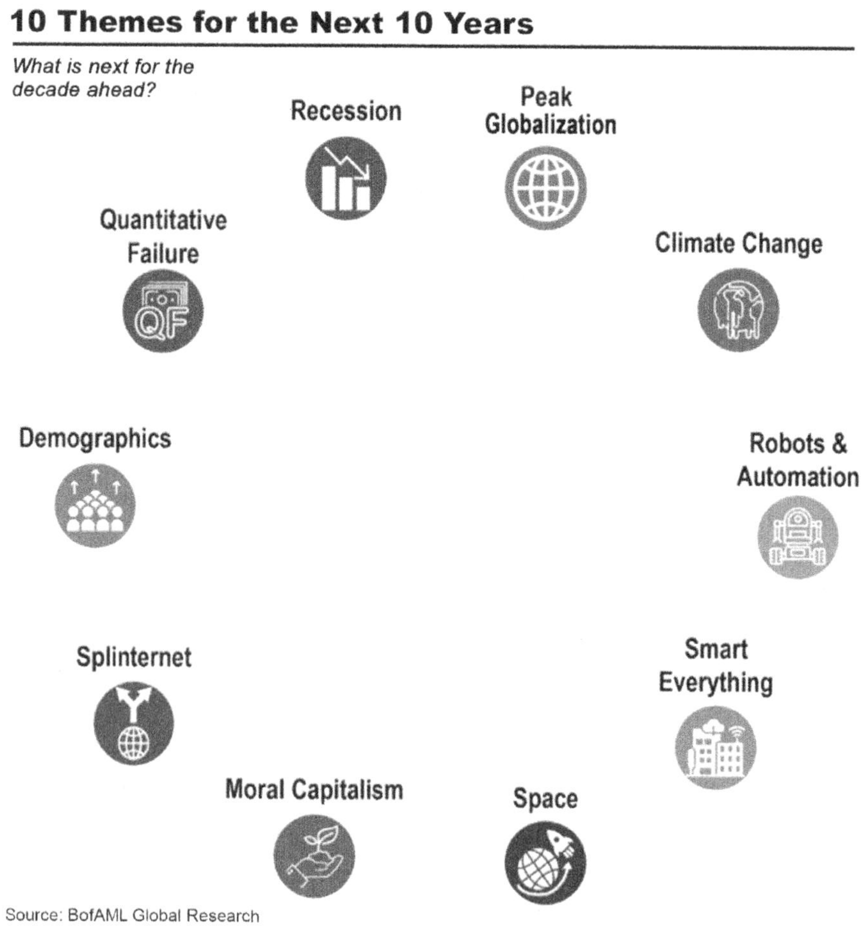

Let's discuss these one by one in each chapter to help you connect further with your education here.

Paradigm One: Peak Globalization

The end of unrestricted free movement of labor, goods, and capital around the world. Winners: local markets, real assets; losers: global markets.

Over the next thirty years, the global population will increase by another billion people. We currently have close to eight billion people living in the world now. That means in thirty years, we'll have nine billion people living on the planet. This will have tremendous consequences.

First, the last time the world increased in population by one billion people, it took one thousand years. This time, we're going to increase by that number in only three decades. What that means is human population growth is exponential. What will that do to our infrastructures?

Think about how humans consume natural resources such as water and energy. How about utilities? And then there are going to be the transportation needs. It will impact the number of devices connected to the Internet, the value of real estate (commercial and residential), and hospitality services. Basically, everything will be impacted by this rapid growth in the human population.

2020 vs. 2030

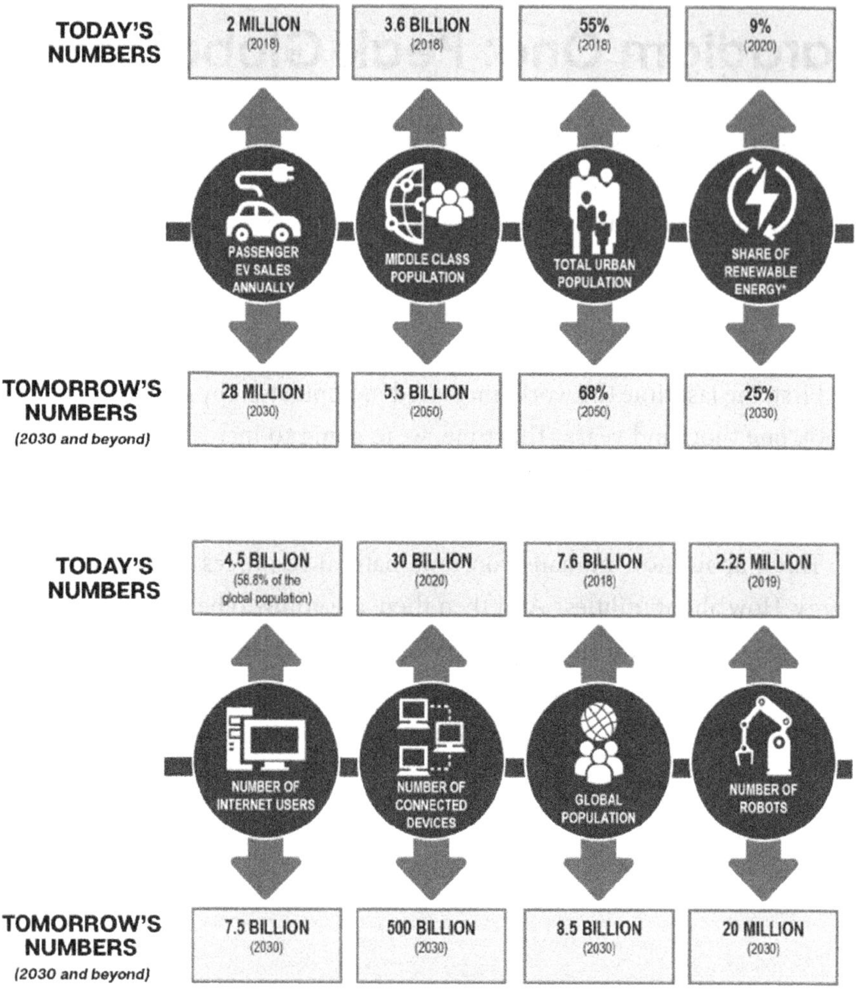

Source: BofAML Global Research, UN, Internet World Stats, Cybersecurity Ventures, Cisco, ITU 2015, Kharas, Brookings Institution, Oxford Economics, BNEF 2019 *variable renewables

But there's one thing that needs to be taken into account very carefully. As technology advances, certain sectors of this technology (robotics, artificial intelligence, etc.) will take over many of the tasks that humans have been doing for a long time. Just because we'll have more people living on the planet doesn't mean that we'll have more jobs for them. What it means is, competition for those jobs will increase. Some industries will experience an employment bottleneck. In those industries, the best competitors, the most educated, the best entrepreneurs, and the billionaires of the world will be the ones to finance future economies.

What are you, as an individual entrepreneur, going to do about that? Are you positioning yourself in the right place for the right opportunity?

The 2020s is the decade of "peak." Most strikingly, we see the 2020s as a groundbreaking "peak decade" that will see many themes reach their inflection points, and business-as-usual investing will be disrupted. Examples include:

Peak globalization: the first time in decades that the unrestricted global flow of people, goods, and capital is no longer guaranteed.

Peak Globalization

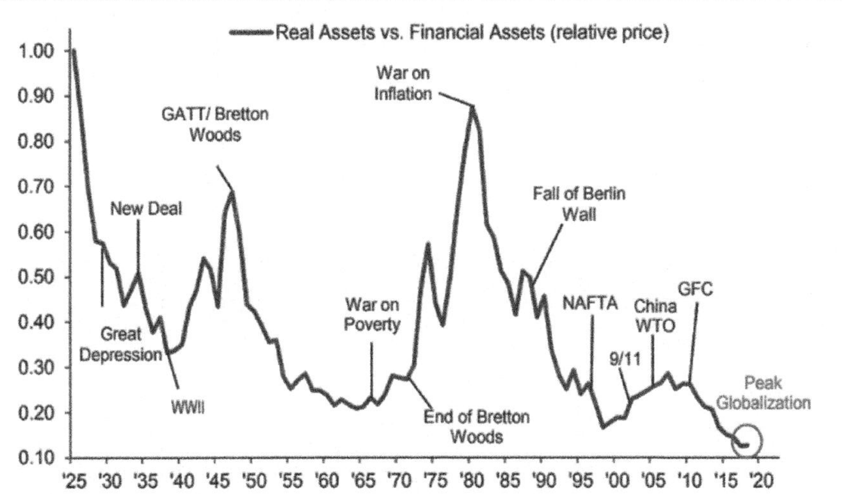

Source: BofA Merrill Lynch Global Investment Strategy, Global Financial Data, Bloomberg, USDA, Savills, Shiller, ONS, Spaenjers, Historic Auto Group. Note: Real Assets (commodities, real estate, collectibles) v.s. Financial Assets (large-cap stocks, long-term government bonds).

Peak inequality: the first time that profit maximization is no longer the sole objective for Wall Street and shareholders, with the rise of "moral capitalism" and the wider importance of stakeholders

Peak youth: the first time that there are more seniors than children in the global population

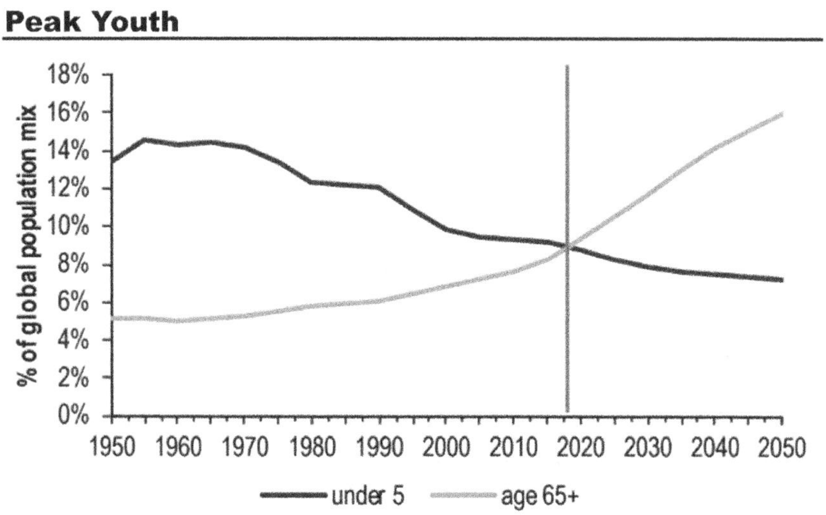

Source: BofA Merrill Lynch Research, UN

Peak oil: the first time that global oil demand plateaus as we transition away from fossil fuels toward renewable energy and electric vehicles (EV)

Peak Oil

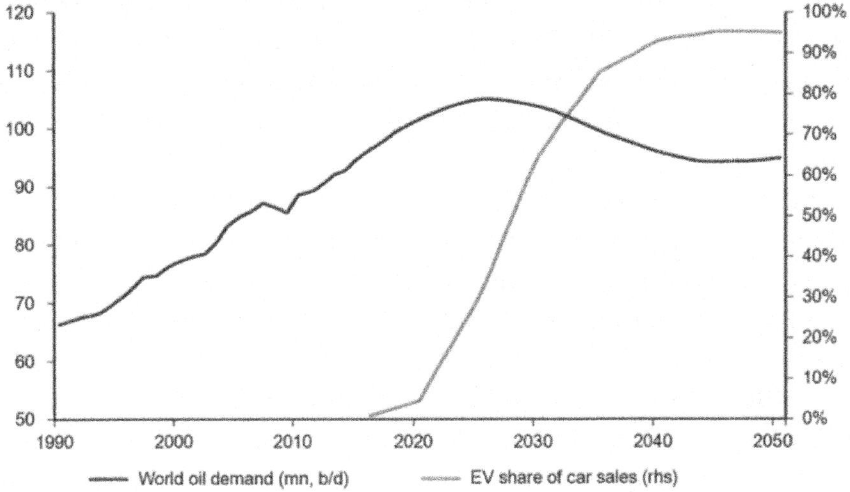

Source: BofA Merrill Lynch Research, IEA

Peak cars: the first time that the total fleet number of diesel internal combustion engine (ICE) vehicles on the road declines

Peak Cars

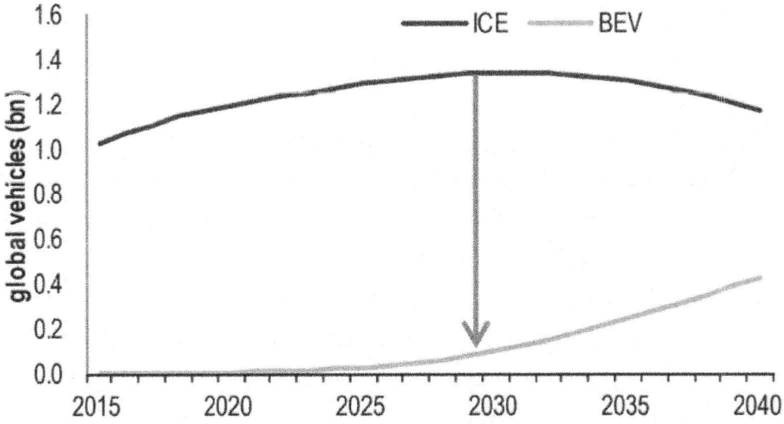

Source: BofA Merrill Lynch Research BNEF // ICE = internal combustion engine; BEV = electric vehicle

Peak stuff: the first time that conspicuous consumption ends in developed markets as we shift away from ownership to the sharing/circular economy.

The "world is not flat" proponents of globalization argue that high inflation in the 1970s was corrected by removing barriers to capital (deregulation, open capital accounts, fall of the Berlin Wall ...). Those same changes are now under threat as societies grapple with inequality and the heavy burden of social security, pensions, and disinflation. The moment of "peak globalization" is likely behind us.

The 1981–2016 era of an unchecked flow of goods, people, and capital is coming to an end, catalyzed by the widespread recognition that while globalization has meant lower consumer prices, it has also meant slower growth, precarious employment, and social disruption. Signs the transition is already under way:

People: After the fall of the Berlin Wall in 1989, fifteen physical barriers were delineating international borders. Now there are seventy-seven. Today, the US is the only large economy to grant unconditional birthright citizenship; other developed countries have skilled or points-based immigration systems.

Goods: The US-China trade war is only the first among many signs that the global flow of goods is being renegotiated. Global trade growth in 2019 (+2.5 percent) is falling below global GDP growth (2.7 percent) for the second time since the financial crisis, a rare event outside of a recession. Bilateral trade battles (US-China, Japan-Korea, US-UK) are replacing multilateral frameworks (NAFTA, TPP), and some corporations are planning relocations of supply chains to regional allies. These tectonic shifts will likely lead to greater macroeconomic volatility.

Capital: From 1960 to 1980, the US averaged a small trade surplus (0.3 percent GDP). Since then, the average current account deficit has fallen to –2.5 percent of GDP. The obverse of any trade deficit is a capital account surplus, which means countries running large trade and budget surpluses (e.g., China, Germany, South Korea) continue to export deflation to the US and other deficit economies. These persistent imbalances generate stronger political will to weaken the US dollar and limit capital inflows, in our view. Local and regional economic ties are likely to accelerate as recent elections (Trump, Brexit, Bolsonaro, Salvini, Orban) proved it is possible to succeed politically while rejecting one or more aspects of globalization.

CHAPTER 3

Digitization of Assets: Nouveau Money

Let's discuss cryptocurrencies in this digital world. Bitcoin made its debut on the heels of the 2009 financial crisis. The few people watching were like, "What is this? A digital currency?" Some of them even mocked the idea.

We all use it. We all need it, some of us more than others. But what is money? Can you define it?

Many people don't understand what money actually is. Where does it come from? Why do we call it "money?"

Gold and silver have been the most common forms of money throughout history. Gold coins were minted in Europe in the thirteenth century. Frederick II is credited with reintroducing gold coins during the crusades. In the fourteenth century, Europe changed from using silver as currency to minting gold. Today, in the twenty-first century, the US Federal Reserve Bank prints money as needed, which is when the market demands it. From time to time, the Fed will decide *now* is the time to print more money.

Before 1976, the US government had an agreement with the central bank that certain gold deposits would back up the currency. In other words, a certain number of ounces of gold would equal a certain amount of US dollars. We'd been operating under the Bretton Woods agreement established in 1944 for monetary exchange, but that changed when President Richard Nixon suspended gold convertibility in 1971. By March 1973, that process was completed when the adjustable peg came to an end. In 1976, under presidents Gerald Ford Jr. and President Jimmy Carter, other nations began to peg their currencies to the US dollar. The Fed just printed more money.

It took some time to adjust to these three administrations. Think about that for a minute. On the ground level, that change created currency inflation. Today, the currency is overinflated.

Apart from being printed and 80 percent of it sent to foreign nations, the US dollar has become the default currency of the world. It's the currency that everybody prefers. The financial crisis happened in 2008 because banks could not sustain the currency valuation without an actual asset, a hard asset, backing it. If you had a million dollars, you didn't really have a million dollars because nothing was backing it, and inflation was eating its value. Bitcoin and several other cryptocurrencies showed up to correct that deficiency, to disrupt the application of money. It allowed people to buy a cryptocurrency and trade it on a secondary market.

The downside to bitcoin at that time was it had no primary market. No one could use it for anything. Today, bitcoin and several other stable digital coins can be spent like money. A company like Soriano Motori Corp can accept bitcoin and other cryptocurrencies as a form of payment in exchange for their goods and services. The reason is that it now has established some credibility. People can trust it.

Since bitcoin was introduced, thousands of other cryptocurrencies have been created. Many of them are traded on multiple exchanges, just like the forex markets, and used to purchase things all over the world. This is happening because people have lost trust in the US government and the central banks. People want to be their own banks.

That's a disruptive market.

This disruption is happening right now before our very eyes. It's happening because the big banks—Bank of America, JPMorgan Chase, Goldman Sachs, and TD Bank, to name a few—can see that cryptocurrencies are empowering people to shift their loyalties to money apps. They can send money with Venmo,

PayPal, WeChat, and different payment systems, and it shows the banks that there is no need for them to act as intermediaries anymore.

The banks freaked out and continued to do so.

They are losing billions of dollars in microtransactions daily. Their clients are leaving and moving away from conventional portfolio management and investments to alternative ones.

They wanted to know how they could get back in the game. Now, these banks are adopting cryptocurrencies and making them a part of their business and wealth management plans to an extent. After the scandalous financial crisis of 2009, many millennials stopped believing in the system. Today, if you are a millennial or a member of Generation Z, you likely have arguments about this at the table with older generation family members. That's an indication that the disruption that took place in 2009 is still happening. The banking industry is still adapting. It's evolving, and bitcoin has revolutionized the financial industry.

The cryptocurrency concept is about secondary market trading, which means you can buy and sell but can't use it to purchase things in the primary markets. The primary markets are the supermarket, toy store, and retailers where people go to buy things and even services. The primary market is necessary because it validates the secondary market.

Secondary market trading is purely speculative. I can buy US dollars or euros based on economic policies because I think one will be stronger than the other. Cryptocurrencies act in similar ways.

When enterprises, companies, small businesses, service providers, and any other primary markets start accepting cryptocurrency as a form of payment, that would validate the secondary market. That's why now is the time that is ripe for this disruption to occur. Analysts are predicting that about 50 percent of the companies on the S&P 500 won't be there in the next 10 years simply because they're not adapting fast enough to the techceleration. That's why disruption is great for some and not so good for others.

The Soriano family legacy started in 1903 with a motor company, a pioneer in pre-war race car manufacturing in southern France.

When World War II broke out, we moved our operations to southern Spain and started manufacturing two-wheeled vehicles, the first motorcycles. Also tricycles, and more. Once again, we were pioneers.

Our family brand wanted a two-wheeled vehicle that could maneuver through forests easily while on their hunting trips; that idea pioneered a new concept for a motorcycle. Spain was neutral in the war, and that allowed our family to gain access to suppliers. Iron was used for armaments for the warring parties. Since Spain was neutral, it did a lot of the manufacturing of armaments for the northern European countries. Our family would buy pots and pans from other Spanish families, melt the iron, and make motorcycle parts.

Another high-demand item for the war was wheels, tires. We'd buy small wheels that were used for very small planes and put them on our motorcycles. When you see pictures of Soriano motorcycles in the 1930s and 1940s, you'll see these very small tires that had come off of airplanes.

My family sold twenty thousand motorcycles then. Amazing!

In 1954, we created our own engines, our own system, and pioneered the industry once again. We influenced Harley Davidson, Ducati, Montesa, Bultaco, and many other worldwide motorcycle manufacturers after that.

Every Soriano generation invested in hospitality. Second- and third-generation Sorianos bought into hotels, hospitals, and pharmacy chains; this became a big market for us, especially in southern Spain, Portugal, France, Italy, and Switzerland to some extent. We were among some of the first hoteliers and got into country clubs, polo clubs, and golf clubs as society evolved into sports and recreation activities. My father himself founded several hospitals, pharmacy chains, and hospitality businesses. We moved away from motor vehicle manufacturing and got into real estate. Then I got involved.

I entered the scene around the same time as the economic downturn. Because of my family legacy and the time that I started getting active in business, I have a different perspective.

By the time I got involved in the business, I had a good understanding of alternative investments, which I learned from my father. Alternative investments are high-risk investments that provide opportunities for high rewards. They often allow the investor to recognize market disruptors. That's what led me to rebirth Soriano Motori—a US corporation with country offices in Italy, Spain, Brazil, Monte Carlo, and Sydney, to date—and start manufacturing motorcycles again. We'll also manufacture scooters, bicycles, and cars, but they will all be electric. The idea is to reduce greenhouse gas emissions.

Alternative investments are trendy. In the US, the most popular wealth management strategy is the nine-to-five job. People go to work every day and put money into a 401K, or some type of IRA. The fund managers invest those funds in stocks and bonds.

The formula for those funds is based on your age. For instance, if you're forty years old, that's how many bonds you should have, percentage-wise, in your portfolio. You should have 40 percent bonds and 60 percent equity. That works for some people, I guess. But how much money can you truly put aside? According to the IRS, an individual can put aside $6,000 a year. The maximum is $7,000 if you are married.

If you start at age twenty and work until you're sixty-five, that's a forty-five-year employment track allowing you to invest $6,000 per year. That's all you can put away for retirement, but is that enough?

Unfortunately, no. Absolutely not.

That means people have to find a better way. They need better alternatives for investing their money so that they get higher returns. You have an advantage if you put your money into disruptive companies because there is always an upside. If you buy at $1 and exit at $10, that's a return of ten times your investment. That won't happen with every investment, of course, but if you work with professionals and understand the markets that you're investing in, you can do pretty well.

Of course, you have to take some chances. That's a big part of our investing approach. You've got to balance your retirement funds in a way that partially has you invested in stocks, bonds, and alternative investments.

That's what the Soriano 8-step manual is all about.

Soriano 8-Step Manual

Everybody has a different concept about their legacy. That's why the first step is to decide what you want your legacy to be.

Step 1: Decide What Legacy You Want to Create

When I ask people what they think a legacy is, some people say, "My name." Others say their legacy is making sure they have enough money, or creating wealth, leaving money for their children and grandchildren. Some people are honest enough to admit they don't know, or they might have the idea that building a school with their name on it or creating a university or institution of higher learning embodies the idea of legacy.

None of those are bad ideas; they're not wrong. But they're not exactly right either because everyone has a different idea about what a legacy is. You have to define yours.

Some people don't care about their legacy. They may not like their surname or family history. They might want to start everything from scratch. Defining your legacy at an early stage gives you a foundation to build on. When you go to talk to investors, presenting your ideas, people will be curious about who you are, where you come from, and why you're doing what you're doing. Your legacy is built from your perspective and that's why it's important.

Step 2: Start Creating Your Legacy Today

Doctors and dentists make a good income in the US. They have disposable income. If you're a doctor, do you want to be a doctor your whole life? Do you want to work forty-five years, contribute to your IRA, and retire, like everyone else, or do you want to quit doctoring in your forties or fifties, sell your practice, and live off of your wealth? Will it be a part of your legacy to leave money behind for your children? Are you interested in being socially responsible, giving money to charities, or starting a foundation? These are the things you'll need to define.

The power of having a legacy is a clear vision for your life. Get it in your head and go with it. The day to start is the day you define your legacy. Don't wait.

Step 3: Simplify Your Life and Focus On the Essentials

It's important to differentiate between the Wall Street guys who are all about the numbers and everyone else. When you get to know someone and what they're about, you can understand why they're successful. You put your money behind them.

As an investor, if your goal is to capitalize, to make money, and that's all you want, there are ways to do that. But I always bet on the person, not the business deal. That's how I approach it.

When you invest in someone, you invest because you trust them. Trust leads to relationships, friendships. The venture may go well or not. Either way, you want to be friends with the individual. They're trusting you with their money, or you're trusting them with theirs.

At the end of the day, do you want to be a billionaire with no friends or reasonably wealthy and with lots of friends? Keep it simple and focus on the things that matter.

Step 4: Treat Everyone You Meet With Kindness

I believe in karma. If you treat people with respect and keep an awareness of what interests them, it will come back around to you. Twice.

Whether it's good or bad, however you treat others will return to you. It may not be immediate. It may be years before you get slammed, but it comes back. When you meet someone and your intentions with them are disrespectful, you're setting yourself up for something bad.

There are a lot of bad things in the world. Fraud. Crime. Bad behavior. Why be a part of that?

As members of a community, society, we should aim at respecting each other. The more people there are on the planet, the more issues we'll have. And there are generational differences, how people perceive the world. It's important to have a moral compass so that you stay on the right path. If you have that, you can do something good for the next generation that comes.

Step 5: Serve to the Best of Your Ability

Serving others doesn't mean you're a servant. It means that you're helping people, that you're pushing them in the right direction.

As an entrepreneur, you want to serve people, a community. When many people do well in business, when corporations do well, they often become more generous. They give to charities and set up foundations. That's a service.

We see slogans all over the place about serving others. For example, the police department says, "We serve and protect." That doesn't mean they're servants, and it doesn't mean they're protectors; they're helping the community. In other words, they're trying to improve the community by ridding it of crime. Unfortunately, there will always be crime. Where there's hunger, thirst, a need, there will be crime. We can choose to serve others or turn our heads and let it be.

The Soriano legacy is built on service to others. When we are successful in a country, we set up foundations to help in some way. That may be through education; we've built schools or built homes for people who didn't have homes. I'm going to continue that legacy by creating my own foundation this year.

One of the ways I plan to do that is to offer scholarships to young adults who want to go to college. That's important because there is a huge gap between the US and Europe in the education that goes along with design engineering. Since my family and I come from both places, I want to be that bridge. I want to help more children get an education and give some of them a full ride to higher education. If they have the talent we're looking for, we'll hire them to work for us, too. Why not?

Step 6: Do the Next Right Thing

This is where your moral compass is at. It's important to understand that the people working for you have families at home. They take time out of their lives to help you obtain success. I'm always conscious of that, and you should be too.

An entrepreneur should be as humane as possible with the people who contribute to their wealth, and the success of the corporation. Doing the right thing means taking care of the people who do things for you. Take care of their families. Get to know them and take an interest in their lives.

If you knew how many corporations are actually doing that with their employees, you'd be shocked. It's not that many. But the most successful companies, in one way or another, have a way to build community within the corporation. It maintains employee loyalty as well as love and respect between management and the employees.

That's what I call doing the right thing.

Step 7: Remind Yourself That You Have Limited Time

Time is gold, so they say. But it's true.

A limited amount of time allows you to push yourself to the limit and beyond. Since there are so many moving parts, to get something done, sometimes you just need to let them run their course. Relax!

It happens a lot with entrepreneurs. They have the best ideas in the world, but they're undisciplined with their time. They end up postponing important events or prolonging the pain of getting started, or worse, they do nothing. Essentially, if this is you, that's precisely what will happen with your business, nothing.

Entrepreneurs must take risks, but they should also be disciplined with their time. Every day.

I wake up every morning at six o'clock. By seven o'clock, I'm in the office. Whether I'm in my office or at home, I'm in my office. My day doesn't end until the end of the day. I take time to enjoy life, embrace a good lunch, have a glass of wine with my friends, and take time to laugh. It's not all about the work, but I'm focused on my time. Even down to a ten-minute nap. It's a part of my agenda to take a nap at the appointed time each day.

The more disciplined you are, the better it is for you to show your investors that you are making something happen. That's what they want to see. Don't waste your time or theirs.

Step 8: Allow for Accessibility

There are two ways to look at accessibility.

First, accessibility to information. You need to be transparent with your investors because they're giving you money. Whether on the phone, social media, text messaging, reports, financial statements, or any information you possess about your corporation, your activities, the information should be readily available. If an investor calls you and says, "I need to take a look at your financials," you shouldn't take more than forty-eight hours to get back to her with the information she is looking for.

The second way to be accessible is to your clients, the consumers. Our approach to our clubhouses is all about accessibility. My customers must be able to book a hotel, reserve a table in my restaurant, book a flight, rent a car, schedule an event, and see the products they are purchasing. Whether it's a motorcycle, an automobile, the latest fashion, or whatever. They should be able to access information and locations for their events.

Make yourself accessible to investors and customers.

Paradigm Two: Recession

A yield curve inversion, global manufacturing recession, trade wars, and rising debt defaults have raised the specter of an imminent global economic recession; I believe the biggest vulnerability for markets in the 2020s comes from today's bond market bubble and responses to the next slowdown will be inflationary.

So what could trigger a recession (bond bubble)?

The combination of globalization, lower interest rates, and a policy and macro backdrop of maximum liquidity and minimal growth have served to exacerbate wealth inequality, and create a surplus of global savings relative to investment. For example, low animal spirits. Higher debt has also failed to deliver higher growth (note corporations and governments are now driven by booms and busts in financial crises).

U.S. Asset Prices and the U.S. Labor Force Participation Rate Since 1978

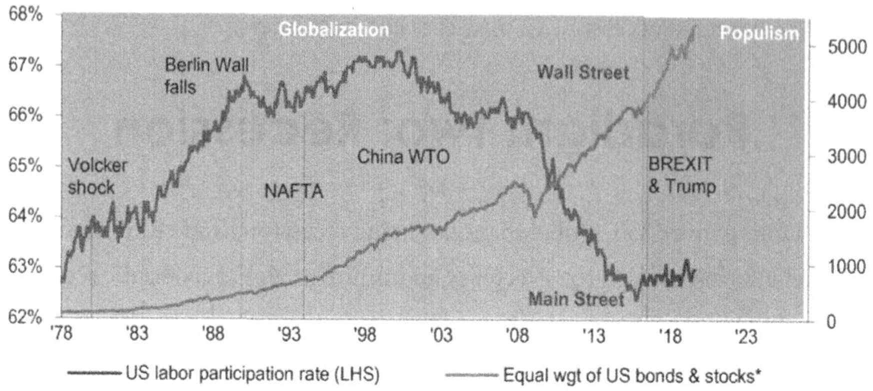

Source: BoA Merrill Lynch Global Investment Strategy, Bloomberg, Haver. *Equal weighted total return index of US equities (DJIA) and US government & corporate bonds (BOAO)

In the coming years, a policy mistake (inflation targeting/MMT) and/or the start of policy impotence (central banks pushing on a string) will likely cause a jump in interest rate volatility, end the decade-long bullish combination of minimum rates and maximum profits and signal the big top in asset prices.

This policy response to the next recession will seek to correct the excesses of the 2010s that led to widespread wealth inequality and financial engineering. These are corporate sector leverage risen by 62 percent, the fact that Wall Street is too big to fail for a record of 5.6x GDP, the private equity and venture capital AUM reaching 4.2 trillion USD on track as a new record for fundraising, share buybacks where US corporations spent 5.4 trillion in shares buybacks in 2019, the shadow banking assets equal to 73 percent of global GDP, and finally OECD zombie companies with an interest coverage ratio below one. Today, post-election, it is quite clear the voting for war on inequality policies involving higher taxes, occupy Silicon Valley policies to target tech profits, regulation of stock buybacks, and a combination of rental price controls. This includes living wages, student debt forgiveness financed by wealth taxation.

CHAPTER 4

Do You Have the Funding?

It's easy to know how much money you have in your bank account. Number one, do you work? Do you have money? Did you save money? Are you an idea originator and now you're looking for capital? If so, what is the budget going to look like? How is the $1 million I give you going to help you get the job done?

All of these metrics need to be put together in the right format. That's usually a business plan, or something similar.

I always say that "I squeeze a quarter coin until the eagle screams." This means I am very cautious with the way investors' money is spent.

If you have funding, do you have it because you saved it and now, you're ready to hit the market and be your own boss? That's usually what entrepreneurs want, to be their own boss. I was always tired of being treated poorly by incompetent people, including my own family at times. I knew I could do better, and I did.

That's all good, but are you putting your money where your mouth is? If so, how much? How much are you investing? What are you doing with that money? It needs to be recorded, and there should be a legal structure in place so that when investors come aboard, they can see that you have your own money invested in the business.

For example, the wealth management world at tier one banks could be extremely attractive to recent university graduates, but do they know that, even in that line of business, entrepreneurial madness is required? You get paid, you get the benefits, but you must know that all that gets reinvested into developing your business with people to manage their money.

In most cases, entrepreneurs don't have a dime to their name; what's the issue there? Investors will have a hard time understanding how disciplined you are with your money when you go from having $10 in your pocket to having $1 million. They want to know if you have good spending habits, or do you act like a rock star with other people's money? That's one of the biggest fears investors have when investing in someone with no funding, or no money of their own.

If you don't have funding and I do, then I have every right to own the idea, pay your salary, and give you some equity in the company. But many entrepreneurs are too proud and won't give up their equity. They want to retain control of their business and my money. Shows like *Shark Tank* in the US and the *Dragon's Den* in the UK are televised examples of that behavior.

This is something you need to decide and define from the beginning. What is your financial position? How much of your company, your freedom, and your ideas are you willing to give up? There are investors out there who just want to take it from you; they're not all nice guys. Most of them have initials that their egos can't fit, and they may be able to do it better than you because they already have the funding. They have the discipline and the networks. However, getting in bed with someone, as we say in this market, means building out that relationship very carefully. Don't ever sell yourself short.

Furthermore, it all comes down to trust, which you establish at the beginning. It is a relationship, and if you have skin in the game, you'll be in a better position. It's important to establish these parameters carefully at the beginning before you get started.

Imagine those romantic relationships where you lied once, then lied again to look good to that person. All of a sudden, it all becomes a big messy breakup. Start out with a true nature and humble approach. I'm not saying that you shouldn't show that you are king of your castle; we want to see confidence.

I always put the first buck in the pot. I make sure the legal structure is taken care of and ensure there is a team in place that I can afford to pay and can work with. No one can wear every hat in the company; that's another mistake entrepreneurs make. They want to wear every single hat and it's not possible.

You will turn gray in no time. It's already happened to me. Sometimes, it's best to relinquish power to those who truly can execute it. I have learned this in my ventures over the years.

Being a multitasker is different from being able to do it all yourself. When entrepreneurs try that approach, they stress out, run out of money, and the business fails. If you're ready to go to market with an idea, make sure you have the basics put together. Go back to financial modeling, the business and marketing plan, your work plan, and a team of people with experience that can advise you and work with you on your idea. That shows your investors that you are not a person who thinks you know it all. Whiteboard it all over your walls. Revisit it and get opinions. Then write out that wonderful business plan.

You do need advisors. You need to show that you can work with people who are experienced in the market. Most of these advisors are people who are either close to retirement or looking to pass on their knowledge. It's similar to the person-to-person transfer of wisdom used by indigenous tribes.

It could be friends of yours, a family friend, or someone you met in college. It could be someone who was referred to you. Then you have to play it nice and respect that wisdom, because it will make you the person you want to become.

When you find these people, offer them some equity, and give them some money for expenses, if required. A quick tip: A typical advisory board gets vested with 1 percent over two to four years, and you pay an hourly rate to compensate for their busy time. If they do not accept these terms, then you may need to find someone else. That's how you build a company from within. By the time you get to the investor, and they meet you and your team, they're going to see true leadership.

To get money on the table, you have to establish trust. You can leverage that position to grow your business even more.

How to Find Funding for Your Idea

There are several ways to find funding today. There's private lending, which is actually quite huge today. But you don't want to pay those percentages. In my father's day, this was called loan sharking. Today, it has become a normal thing.

There's also the convertible note approach, which is essentially taking a loan from various investors and, in return, offering them a return each year for the duration of the note or converting the note to equity when the right time comes for them. You may want to include a buyout clause so that they are happy with an exit. Anything more than whatever today's markets' returns offer is already enough.

The right time is when you're ready to sell a product or service and you have revenue. Then you can present the equity option to the investor. You ask them, "Do you want to be my equity partner or just pay the loan off?" Those that believe in you will always take the full ride and share the risk and interest with you. Those looking for some extra money will cash out.

Either way, that's a decision that will need to be made. During the time of the note, as you're working with the investor, you'll get to know them, and they'll get to know you. You'll find out if they're there to help your company grow or if they just see themselves as a lender. It is your job to bring on board smart money so that your company's strategies make revenue and profits.

There are also crowdfunding platforms where you can get money from multiple investors. US law allows you to raise up to $5 million through crowdfunding, and that's a chunk of change if you have a good idea. I'm skeptical of any business that needs more than $5 million to get started, unless you're building a rocket ship that is going to Mars. I'm definitely going to be the first passenger on it.

When asking investors for money, you have to be careful how you sell your idea because investors know the numbers better than you think. They know how much things cost, how to leverage, and how to make it happen. In the beginning, you should outsource as much as possible to save yourself some money and headaches.

Again, try not to reinvent the wheel. There are loads of tech companies that can help you run your business in a smart way cost-efficiently. From cloud services to email management, get advice from multiple professionals. When you go through the exercise of looking for funds, offer some equity to family and friends first in exchange for their risk. Give them the first shot at getting involved. At times, you can get money from your folks at a zero-interest rate, or zero equity. That is the whole point.

To get big money from investors, you need to be fully prepared with a serious business structure in place. You should also have some revenue, which doesn't happen very often with start-ups. So grow organically and, God, try not to use Wall Street metrics to evaluate your firm. Anything above 10MM on day one is already a loser.

You can probably think of multiple entrepreneurs who have been successful at more than one venture in their lifetimes. Make a list of people you can go to. Have a good lawyer help set up the organization and a good accountant who can keep your finances straight. You're going to need those professionals for the life of your business. Make friends with those guys. They are necessary at all costs.

In terms of time frame, there's no optimal period of time for when you start raising funds. It depends on the industry and your business plan. You will always be selling and raising money, regardless. Start all the conversations, join forums, conferences, and invite them all. Put together something that differentiates you from all others raising capital. If you read carefully all the paradigms of this decade, you can see how much money is moving around daily in ventures worldwide. Everyone wants to be a part of the next best exciting thing and accumulate wealth.

In 2019, I spent the last six months of the year researching my market. I wrote a business plan more than ten times. I played around with financial modeling, trying to understand whether I had the wind to my back. By 2020, I presented my idea to a group of friends and family and every single one of them put money into my business. At that point, I knew I had something good going on. I put more money, time, and brainpower into it myself. Today, a year after the pandemic, I have a number of proven concepts and a proven business plan with multiple revenue model integrations and market segmentations. We're now selling the product and expanding out to dynamic verticals that will complement the brand. These include clubhouses, fashion brands, futuristic car designs, drones, motorcycles, scooters, and boats, all in the EV, AV, and AI realm.

One idea snowballed into a whole lot more. It's amazing. Now investors are calling me to invest in my company and don't even know me. I have to make a decision on whether to accept their money or not; that is the position you want to be in. When you go to a supermarket to buy your fruits and vegetables, you always want the freshest ones with the appeal to your health and palate. That is exactly what investors shop for.

But you do want to select the right investors carefully. Whose money do you want in your company? Who do you want to write your success story with? Not every dollar is a good dollar. Remember that every second of every day.

How Much Equity Should You Sell?

I use a straightforward formula.

Your company valuation, initially, should be below $10 million. If a start-up comes to me with a business plan, that shows they are way over $10 million valuation pre-money. They likely have no revenue, and I just say to them, "Have a nice day; it was great meeting you."

It's better to be below $5 million. $3 million is better. $2 million is even more feasible and makes sense. How many people do you know running multimillion-dollar businesses? How many do you know running a $1 million business? Most business owners are at the shopping mall looking for that kind of money.

The formula is simple. If your pre-money valuation is $10 million and you want $1 million of my money, then I should have 9.09 percent of your company.

The lower the valuation, the more equity I can take for the same investment. For example, if you say you're going to make $2 million and you want a $1 million investment from me, that's 33.33 percent equity.

Numbers are numbers. Don't get too creative with it, or go get a loan.

When it comes to people working for you, it's called sweat equity. Ask the person how much they think they can make each year in salary. Take the salary and multiply it by five years. Divide it by the time that you expect them to work with you. For instance, if a salary is $100,000 a year for a web developer, multiplied by five years, it comes to $500,000. If we agree you'll work on my project for two years and you could expect to earn a $100,000 per year salary, then your sweat equity is $200,000. Your equity will be the equivalent of that income. Be nice to your workers; they are the engines of your cars.

As the structure of the company is developed, you'll be able to actually earn your salary. By the way, start a payroll system as soon as possible so your IRS filings are solid. There are multiple systems now that are extremely economical. Forget ADP with your bank; old and expensive is not needed.

Of course, there are multiple ways to figure out how much equity to share. If an entrepreneur says they have a $10 million company and need $1 million to keep it going but offers me just 1 percent in equity, I'll walk away from that. Clearly, they don't understand mathematics.

First, that's a sign of disrespect to the investor. If you want smart money, it comes with a price. If you want dumb money, then it's readily available around the block from where you live. But you have to know your math, what you're offering, and what you can get accomplished. If you say you're going to make X amount of money and you fall super short, that doesn't look good. Then I'll have to buy you out, replace you or your management, or figure out another solution.

Entrepreneurs are not always CEOs. You may be the founder of your corporation but not the chief executive officer, or any executive. Maybe you need to sit it out and let the professionals run your business with your vision. It does happen, and I've fired entrepreneurs before to hire the right people. You may need to hire a CEO to help you achieve your goals. They have talent, they have knowledge, and they have networks. Most importantly, they have experience attached to wisdom.

When I invest in a new business, I like to know that the founders know they are not experts. I like to see that they're going to bring in the experts or that they expect me to help them find the experts and invite them to the table, too. If their ego or pride gets in the way, then I walk away.

Being a founder is just as important as being CEO. For instance, with Soriano Motori, I'm the founder and chairman. That means I still have control of the company. I own the company, and it operates on my vision.

The CEO is my second hand, the operational guy. He understands logistics. I can put him in front of a client and the client will call and say, "I like him; he's a nice guy. I'm going to do business with your company." Imagine being the president with several generals for your armed forces; that is you.

It's important to identify your role and the role of the experts you bring in to help you with your business. Define all of that up front because you need to have the right talent. Write it down and have an MOU in place so that accountability is measured.

Some people prefer to deal directly with the founder, not the CEO. That's fine. The founder needs to be accessible to them. Maybe an investor wants to talk to the founder. They might want to speak to the person who envisioned the whole thing. They might say, "I need to understand why they think this idea is going to disrupt the market." They have that right, so make yourself accessible.

There are people, small investors, who will be proud to put their money into your business. They really like what you're doing. They're proud to tell their friends that they've got their money invested in someone like you. Respect that, embrace it, and make yourself accessible.

Paradigm Three: Quantitative Failure

Investors are concerned that more monetary policy alone will not bolster animal spirits or deliver meaningful reflation despite 768 interest cuts and +12 trillion USD in quantitative easing since 2009; banks are at risk from negative interest rate policy (NIRP).

This is clearly a failure, or monetary impotence, because it diminishes returns from increasingly activist central banks and their use of these policies to regulate our economy. Since the financial crisis, central banks have come up with similar old recipes such as QE, yield curve controls, zero interest rates, NIRP, and excess liquidity. I know these things sound foreign to you but keep on reading until it connects.

The excess supply of money has been outweighed by the four deflationary Ds of aging: demographics, excess debt, bank deleveraging, and technological disruption.

Worst of all, rather than stimulating reflation, monetary policy gains are locked within the financial system driving inequality on Wall Street and Main Street. This means, in turn, that we should be concerned with another recession, a bubble in bonds, and the rise of populism.

Monetary policy measures are proving less and less effective at boosting corporate and household "animal spirits." Winners: Keynesianism, gold; losers: financial assets, monetarism.

Quantitative failure, or monetary policy impotence, refers to diminishing returns from increasingly activist central banks and their use of monetary policies to reflate the economy. As our European credit strategist shows, a net 33 percent of credit investors say that central banks' bare monetary cupboards are their biggest fear.

Since the global financial crisis, central banks have pursued a series of extraordinary monetary policies to bolster inflation and growth: quantitative easing, yield curve controls, zero interest rate policies, negative interest rate policies, and excess liquidity. The startling combination of 768 interest rate cuts (1 for every 4 trading days since March 2009) and $12.4 trillion purchases has taken interest rates to five-thousand-year lows (chart 2).

However, real GDP has averaged a meager 2 percent in the US, and 1 percent in the EU and Japan over the past ten years, with inflation equally subdued. The excess supply of money has been outweighed by the four deflationary "Ds" of aging demographics, excess debt, bank deleveraging, technological disruption. Tempered "animal spirits" have also been compounded by a surplus of global savings glut relative to investment.

Rather than stimulating reflation, monetary policy gains have been locked within the financial system driving inequality on Wall Street and Main Street. This, in turn, is leading to concerns about recession, a bubble in bonds, and the rise of populism—as addressed previously.

Pushing on a String

We entered the 2020s with central banks already close to or below the zero bound: Federal Reserve Bank, 175 basis points; European Central Bank, 50 basis points; and Bank of Japan, 10 basis points. The risk is that, in trying to fight an economic slowdown, central banks either run up against their own political constraints and red lines or start pursuing increasingly extreme measures that provoke counterintuitive and counterproductive responses.

If central banks were "pushing on a string," monetary policy alone would not stave off recession risks or populism. Instead of easing financial conditions, central bank interest rate cuts compounded issues. Extreme policies simply drove precautionary saving and defensive behavior by economic agents.

> "Be fearful when others are greedy and be greedy when others are fearful ..." — Warren Buffett

CHAPTER 5

Putting Together the Idea

The world changed drastically from BC to AC (before corona and after corona). As I write this in 2021:

- All G20 countries have implemented some sort of shelter-in-place order, equivalent to 80 percent of global GDP under lockdown.[1]
- Global carbon emissions could fall by 8 percent this year, equivalent to what India emits per year, which would be the biggest drop ever recorded.[2]
- The fall in greenhouse gases this year alone could be greater than all the recessions of the past fifty years combined.[2]
- The Fed was buying more than $1 million in financial assets every second during the peak panic phase of the crisis.[3]
- It took airlines sixty-four years to reach fifty million users, and Netflix seven years, but Disney+ achieved this in just five months with the kids under lockdown (Gen Z bonus stat: it took the HouseParty app just 1 month!).[4]
- At four billion people, more citizens are now under lockdown than those who have access to Internet broadband, social media or even safe toilet sanitation.[4]

- The US $20 trillion market cap of thematic enablers identified in the post-COVID era is bigger than the GDP of the US, Europe or even China plus India.[4]
- About 90 percent of painkillers and antibiotics (ibuprofen, penicillin, etc.) have active drug ingredients imported from China.[4]
- For the first time in history, China is outspending the US on R&D ($501 billion vs. $493 billion in USD).
- The number of unemployed Americans now exceeds the population of Australia.[4]
- More US jobs have been wiped out in the last month than were created in the past ten years since the Great Recession.[5]
- Sir Isaac Newton discovered gravity during the Great Plague of 1666 when he had to practice social distancing and work from home.[6]
- Zoom daily active users jumped thirtyfold in just four months from ten million in December to three hundred million in April 2020.[7]
- Zoom's terms and conditions are longer than the US Constitution.[7]
- Fortnite is now more popular than football based on Google searches.[8]
- If you were to stack up the $2 trillion US fiscal stimulus package in one-dollar bills, they would be 6x higher than the International Space Station (about 1,500 miles).[4]
- If you were to put them side by side, they would stretch from Earth to Mars … and halfway back again![4]

Sources: [1]Bond Capital, [2]IEA, BofA Research, Guardian, Carbon Brief, [3]BofA Research, FT, [4]BofA Research, [5]New York Times, [6]Bloomberg, [7]Zoom, [8]Google Trends

As an entrepreneur, you've got to have a line of sight. Put your road map in front of you. It shows you have discipline. It shows investors that you have vision and ambition. It shows the true spirit of entrepreneurism because you're laying out your blueprint. You're planning your success. These are the key elements of a disruptive business idea.

A great perspective is to know and understand the year the earth stood still. Yes, 2020, that is. What a year for all of us? Keep in mind that similar to other historical global crises, COVID-19's influence will and has been extending beyond the current lockdown peaks. This pandemic is accelerating many macro trends that would have taken at least five or more years to play out before, from peak globalization to renewed tech wars and a reappraisal of health care systems and governmental influence.

There are five primers, that as you read and put together your ideas in sight, should be considered due to the aftermath of COVID-19.

1. Geopolitics and globalization: global to local supply chain diversification and climate change corporation as seen in the whole automotive industry.
2. Tech wars. This is more along the lines of all sorts of digital transformation acceleration with long-term enablers such as big data, AI, 5G, and cybersecurity. Yes, keep in mind the lockdown tech we are all living with working from home, e-commerce, and moon-shot technologies.
3. Big governments. Right! The safety over privacy discussions where there is a stakeholders-versus-shareholders approach. And of course, our personal safety boom.
4. Health. That is the social focus on ESG for all business activities, public health, health-tech (just as you are measuring the calories of all that you eat and whether or not you are taking the steps you need), in some places even the talk that health care is a national resource and government intervention all around.
5. The new consumer: so far Gen Z leads the way and inequality at work got polarized by COVID-19.

That said, your business plan should be as clear to others as it is to you. If an investor walks away from a meeting wondering what the entrepreneur was talking about, then you've failed to communicate your vision. You've failed to begin a relationship, and you've failed to show how you're going to disrupt the market.

When you walk away from your presentation, you want people pulling their hair out, asking themselves, "Why didn't I come up with this idea?" You want them backing out of the room thinking they've been wasting their time by not seeing the opportunity. That's the effect you want. Most importantly, understand your environment.

Even if an investor isn't good for your business, if they walk away spreading the word about your idea to everyone they meet, then you know you've hit all the right buttons. They should run to their friends screaming, "Holy cow! I spoke with this guy. He's amazing. Check out this idea."

Yes, this brings me to NDAs or nondisclosure agreements. During this phase, unless you got some serious patents, trademarks, or copyrights, then why are you paranoid about it? You need the money. But do feel it out. If your Spiderman senses tickle, walk away. Find a better person to finance your idea.

It shouldn't take more than thirty seconds to describe your idea. Forty-five seconds, tops. Unless it involves brain surgery. It should be quick and sharp. It should trigger ideas in the minds of your potential investors. Spend more time on the details and less time describing your idea. The journey of an idea begins with the description, but it doesn't end there.

Become a Person Who Reads

People today don't read much, but I think they should. Entrepreneurs should read at least one book every two weeks. People spend too much time watching television and searching the Internet. Instead, read poetry and literature. Learn how to communicate. How to write. How to connect with others' interests.

I visited Key West, one of the places Ernest Hemingway lived for eight years of his life. He kept the house until his death. When I was there, I wondered, "Why does a guy like him, a great writer, come to a place like this?"

When I studied his environment, I realized that there were elements of inspiration embedded within it. I could understand why he found himself there and was able to write some amazing books.

Hemingway had the mindset. He set out to be creative and to communicate eloquently to a crowd. Which is not easy, it requires training. In other words, communicating an idea to a group of investors is the hardest thing you'll ever do.

You can learn to give better presentations by reading other people's presentations. The poets. Hemingway. And speeches given at the United Nations by heads of state. Read their speeches. Watch them on cable TV. Learn from them. See how they are structured and learn how the speechmakers deliver their addresses. You want to plant your idea into as many heads as possible, and to do that, you need to learn how to structure a business plan, organize it, and deliver it to the right audience. If you do that well, people will talk about you regardless. All this is organic. You want this advertising; we will discuss what that means as well.

Study the journey of an idea. Learn how it develops and how it changes over the course of its lifetime. If you can do that, you'll have greater success.

You don't want people saying, "Forget it; this is a waste of time." Rather, you want them saying, "I want to talk to this guy again, invite him over for dinner." You want people to invite you to their house, meet their children, and influence their entire family. If you can create that kind of effect, you'll have a winning idea, and it bears a chance of being disruptive. I've had that effect on investors, partners, and members of my family all of my life. It begins with learning how to read books.

The Journey of an Idea

I call the process of building a company from the ground up "the journey of an idea." It starts in your head and journeys through time. It travels from your mind through the air to your team and becomes a production process that delivers a product in the real world. It's the journey of an idea.

Keep in mind that there is a race to supremacy to spark a new wave of investments. This was presented by COVID-19 while accelerating structural shifts in consumption and working, and our use of deployment of technology. The lasting legacy of this will be the combination of a new communication infrastructure, data generation, cloud computing power, and bandwidth.

But let's keep it a bit easier to understand how to even begin the process. In terms of how far out to go with your business plan, three to five years is normal. But you need to plan your exit. You should be thinking about how you're going to get out before you start. Based on your exit strategy, create a road map that makes sense. Every quarter, every month, every week, have a work plan so that you can lay it all out and execute your deliverables. I am, for one, a big fan of whiteboarding a whole year on the biggest whiteboard monthly. Then share that with my team and hear their feedback. We then revisit the initial whiteboard framework and see what we are still missing.

When you work with a team, either you're a good project manager or you hire someone who is going to keep track of all of your objectives and report back to you. I start my week by putting all new ideas on the table first thing on Monday mornings. At times, Sunday evenings to get everyone ready to go the next day.

It begins with an open door. We go through the week and try to get an understanding of what everybody is going to be doing that week. Friday afternoon, or the next Monday, we review what was accomplished and what wasn't accomplished. If something didn't get done, I want to know why. You learn from your failures, from your team, and figure out your strengths and weaknesses.

As a founder and owner, you should be aware of everybody's personal problems and working position within your company. What motivates them? What personal factors motivate them to come to work and perform? These are the things that are going to affect your vision, your deliverables, your expectations, and your personal and corporate actions.

Your road map has to be very detailed. Whiteboard it every week, as I do. I get with my team and we draw it out, either virtually or physically. Doing that gives everyone a perspective, a horizon to shoot for in accomplishing their piece of the organizational puzzle. We get things accomplished by keeping it in front of us all the time.

This is very important; I get the whole team together. I also get together with my core team and do the same thing. Once a month and once a quarter. We present a quarterly report for all investors, team members, and professionals. That includes lawyers and accountants involved in the business because I need them to know what I'm thinking and what I'm trying to do.

Sometimes, I invite families of investors or team members because they need to know what their husband or wife is doing by investing in our ventures. It helps them establish a purpose for their lives, too. We all need purpose.

Everything starts at home. If it's emotional, it begins at home. If you don't communicate with your husband or wife about what you're doing in life, what motivates you, what brings money through the door and puts food on the table, they can't support you. You have to have a strong, emotional and psychological bond with everyone to build a successful organization.

It starts at home and moves into the office. There, you can see how it manifests with your employees, other leaders, managers, and the whole team. From the bottom to the top. If you're aware of that, you'll run a healthy organization.

An idea travels. It starts at home and goes out from there. Your vision is connected to your ability to communicate your ideas to the people in your life who can help you make them happen. When everyone is aligned with your vision, you have less friction, fewer problems, and communication breakdowns happen less frequently.

When you live this way and you meet strangers, by the time you meet them, they'll already know something about you and your company. You already have a foot in the door with investors because someone has told them about you before you shook their hand.

It's called leverage. It happens everywhere you make a connection, even social media.

Social media helps you communicate what you're doing. Use the tools at your disposal to facilitate the journey of your idea. Once it's out there, you want people to talk to you about it. That's how you attract new clients, investors, friends, and others who will be impacted by your idea.

Underlying infrastructure may differ regionally, though, due to ongoing trade tensions, which technology is at the heart of. This could pull forward the commercialization of moon-shot technologies such as autonomous vehicles, quantum computing, and vertical farming, as nations look to get an edge in the race for tech supremacy and shifting industrial and political priorities.

Nearer term, trends like working from home, e-commerce, and stay-at-home activities (such as streaming and e-sports) could see increasing long-term adoption. However, there are several short-term challenges. Economic uncertainty may constrain overall demand and funding.

When people are inspired by your vision, they become your advocate, your salesperson, your partner.

If you think about the most successful companies in the world, people talk about them as if they owned the company. When you buy their product, you talk about it as if it's yours. When you buy a Louis Vuitton bag, an Apple TV, or a Samsung phone, you embrace it as your own. That's what you want with your idea. You want people to become ambassadors of your brand, your vision, and your idea.

Milestones show investors, clients, and others what you've accomplished. For example, during the pandemic last year, we started drawing images on napkins of what the Soriano motorcycle would look like; we called it "Drawing out the lines." From the napkin, we took the drawing to a piece of paper and had someone with technical design expertise draw it. Then we mapped it out.

Once we had the design, we needed to understand how the parts of the motorcycle fit into the design. It all had to make sense: the symmetry, the weight distribution, aerodynamics, and the aesthetics. We put our brainpower to work. Between engineers and designers, we pulled it off.

If you follow the pattern, you'll have a drawing before you know it. You get your schematics and head off to the suppliers to build the product. They will literally build all of the parts for you. They'll also advise you on what raw materials to use.

As you work with suppliers on the different parts and the materials for those parts, you'll need to learn the weight of each part. Remember back to elementary school when you learned the different elements in your chemistry class with that boring low-voiced teacher. Each piece of my motorcycle has a weight based on the raw materials used for the engines, the chassis, its size, and dimensions of each part.

Working with your suppliers, you'll learn all the minute details that will become key communication factors when speaking to investors. You will also learn that each supplier is extremely proud of what they do, even if it's just a bolt or nut for your mechanical ideas. Appreciate the whole process.

Investors may not be interested in every single detail, but you are the expert and knowing those details shows that you know what you're doing. You want to leave the room saying, "I know what I'm talking about." You're prepared for every question, and you know the answers. You educated them.

When you're the expert in the room, investors will leave and talk about you. They might not get the idea or understand your vision entirely. But they will listen to it and tell people they know who might be interested. That's how you connect the dots.

We started this process at Soriano Motori in 2019 for the rebirth of a strong legacy brand in Europe. After eleven months, we had assembled a motorcycle 100 percent for all to see. This has never been done. We could mount it, start it, and ride it. That was an industry record-setting time. We did it because we were dedicated to making it work.

Now, we're in our second year. We have sketches of scooters that we're going to bring to market. We'll present those at the end of the year. We started designing our own fashion collection to support the brand and its elegance.

The same thing goes with our fashion line and private social clubs. We're determined to draw it out on a napkin, hire a technical designer, take it to paper, and go to print. If your idea travels from the germ stage to an actual physical deliverable, then you've proven you know what you're doing and that's what investors want to see. They're not just interested in the idea. It's not the words that fly through the air that are important. It's the intangibles, the revenues. You want to show those too, as soon as possible. If you have them, they will shut up your critics. It will shock them first; then they'll envy your market share. They'll want your consumers, your share of the market.

Do remember that concerns over social distancing may negatively affect sharing economy platforms. Increased data tracking and personal monitoring raise privacy and ethical concerns that will vary internationally. Beneficiaries: cloud, 5G, processing power, robotics, autonomous vehicles; challenged: privacy, old media, bricks and mortar, sharing economy.

Plan Your Budget, and Stick to It

People say, "You need money to make money." That may be true, but the entrepreneur must get in the habit of underspending and overproducing.

If you buy a television set from one hundred different outlets, do you take the time to understand which one of those outlets will give you a better price? Price comparison is not cheap. It means you're being practical with your money because you want to reduce expenses. If that's the kind of person you are, it means you're going to look for the best ways to reduce the cost of legitimate purchases.

Reducing costs maximizes profits. You're looking for that type of manager. As an investor, you should always have that mindset. As an entrepreneur, you must train yourself to be that kind of business manager.

You might say, "These big tech companies spend billions of dollars every month." I say you're not a tech company. At the start-up stage, you're just an idea. What you need to do is translate that idea into something tangible, bring in some revenue. After you do that, then you can call yourself a tech company.

An idea with no clients is nothing; it has no value. First, convert clients. If you don't, you'll have zero value. Perpetually.

Investors don't like to see companies waste money. You might as well go to Vegas and gamble it all away on chance games. That's essentially what you're doing with it when you don't count your costs. Remember we spoke about accountability. Well, that is you at all times.

You have to have a budget, that's very important. Your budget is the bar you've set on what can be spent for your company. Every dollar you earn has to translate into money spent. If you spend a dollar, you should get two back. You break even; then you profit a dollar. That should be every entrepreneur's standard.

Some companies take longer to profit because they go through expansion periods. Airbnb and Tesla come to mind. They spend billions every month for this and that. When are they going to be profitable? Profit is the only true metric. And investors will always ask, "When will you be profitable?" That's a question I always ask. Some economists called the stock market the barometer of our economy. Wrong. Our economy is truly measured at a small to medium private business scale. That is what we do when we visit the local bagel shop in our neighborhood.

Your budget is your reflection point. That's the metric you use to determine your success. It helps you make financial decisions. As an investor, my due diligence takes me deep into a business's financial statements. I want to see the entrepreneur's financials. What do their taxes look like, and so on? I do that because it's a part of the system. If I don't know how you manage your own money, whether or not you're responsible with your taxes, then how can I be confident that you can run a multimillion-dollar company using my money?

That's why I always go deep. I want to understand whether someone is financially responsible or not. If I'm going to gamble on somebody with no experience, then I may as well give to charity and collect a tax incentive from the government for doing so.

The budget is elementary. It's a reflection of how an entrepreneur manages money.

Another side note just to keep in the back of your mind as the journey is taking place is to know who you are targeting. Gen Z may be uniquely prepared for the new era of interactions following the coronavirus crisis—other generations will need to quickly learn from them and change their consumer habits, enlarging the addressable markets for technology products such as streaming, social media, e-commerce, and so on. On the flip side, this generation—alongside millennials—is the most exposed to reduced earning potential over the long term. Further, this global pandemic could result in a slowdown or reversal of the decades-long march to pull billions out of poverty—just a 20 percent fall in income could push over half a billion people into poverty.

Finally, following a fall in births nine months after the crisis, there could be a baby boom, as seen after many famines, earthquakes, and disease outbreaks. Beneficiaries: Gen Z, new media, payments, ESG; challenged: millennials, bottom billions, traditional consumer.

Paradigm Four: Climate Change

We are reaching the inflection point in climate change; that is, by the end of the decade, humans could exhaust the remaining carbon budget for a two-thirds chance of remaining below 1.5°C warming. Global sea levels could increase further, economic damage from extreme weather will rise and the economic/

political damage could result in climate-induced immigration, inequality, and unemployment.

At the same time, the 2020s will be the decade of clean energy solutions and bold action could boost the global economy in a "make earth green again" (MEGA) strategy.

Historical Trends of Climate Change

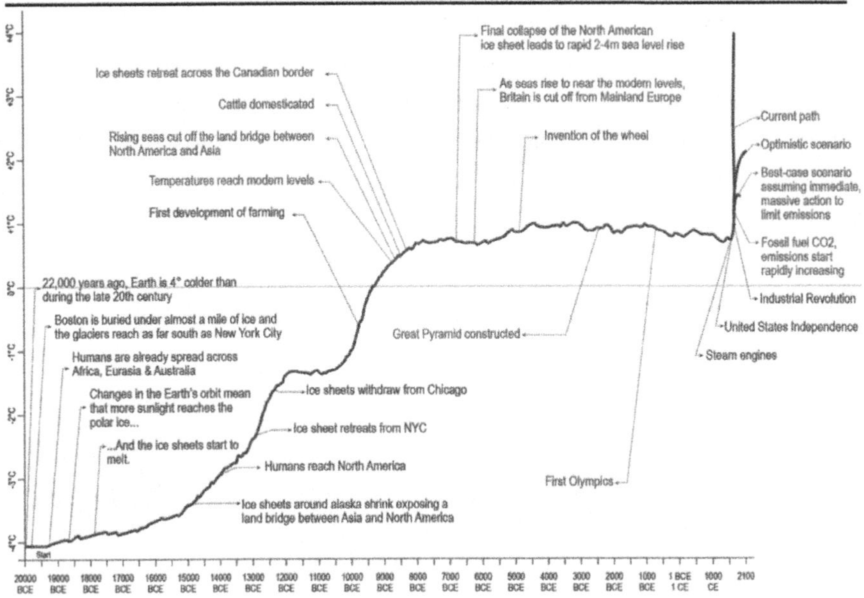

Source: BofAML Global Research

No longer business as usual, climate change is disrupting the way we live. Accelerating global warming means the past four years have been the hottest ever recorded and the twenty warmest years occurred over the past twenty-two years. More extreme weather episodes are leading to large-scale displacements; from 2008 to 2015, more than twenty-five million people on average were internally displaced each year.

Meanwhile, the global reliance on nonrecyclable plastics is contributing to more than eight million tons of plastic waste hitting our oceans each year, affecting fisheries and marine ecosystems (source: Ellen MacArthur).

The two biggest emitters are China (29 percent) and the US (14 percent), where the cost to the US economy is already estimated to be US$250 billion per annum. Inequality of use and access to resources carry developmental implications too: Luxembourg's ecological footprint is a startling 31x that of Eritrea's (source: Global Carbon Atlas).

Climate action could be one of the biggest transformations to the global economic landscape, impacting current practices in multiple areas, such as energy, farming, transport, and even globalization (local production can be much more environmentally friendly, for example). Bold climate action could yield a direct economic gain of an estimated US$26 trillion through 2030 compared with business as usual (source: New Climate Economy).

Regulations around emissions and waste are likely to become tougher and tougher—there are many signs of this, such as 100 percent renewable targets, carbon pricing laws, proposed ICE bans, higher city charges for polluting vehicles, airline emission taxes, and so on.

Fossil Fuels vs. Renewables

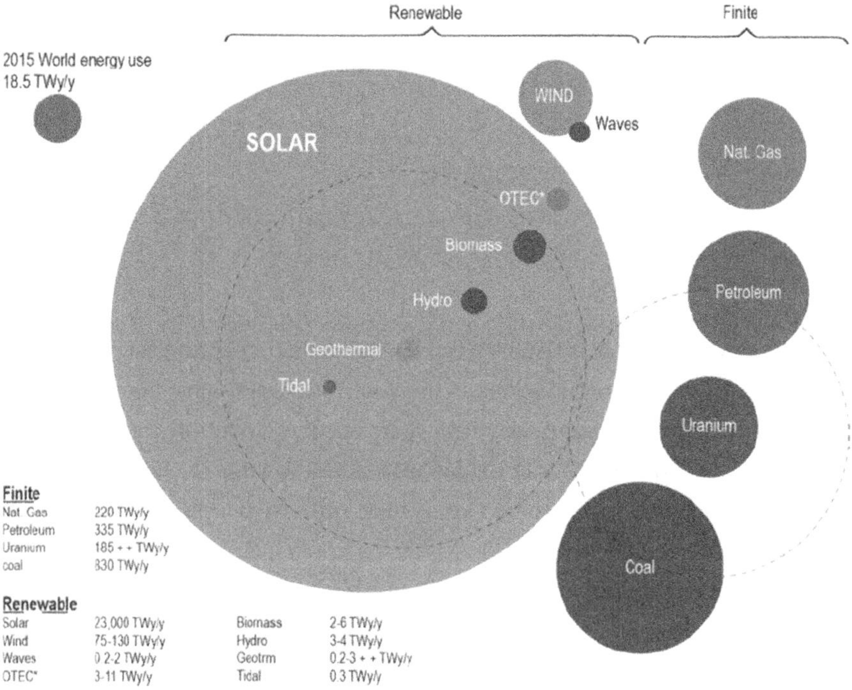

Source: Perez & Perez, IEA 2015

It's a decade for action. The UN predicts that the 2020s will be critical to limiting global warming to the 1.5°C preindustrial level by 2030—the temperature above which academics believe climate change will cause lasting environmental damage. A rise of just 4°C would be enough to submerge six hundred million homes due to rising sea levels.

One market-based mechanism for reducing emissions is carbon pricing. More than forty governments worldwide either directly tax fossil fuels or use cap-and-trade programs, with the value of carbon trading surpassing US$40 billion in 2017. Such changes—for example, new EU legislation tripling CO2 prices—incentivize energy-intensive industries to embrace more sustainable long-term energy sources as renewables become more price competitive.

Efforts to curb global warming require behavioral and systemic changes that will provide substantial opportunities for investors. For example, New Climate Economy predicts bold climate action could yield a direct economic gain of US$26 trillion through to 2030 compared with business as usual. Tougher regulations around emissions and waste management, as well as carbon targets, will also change how investment decisions are made.

New technologies are also driving change within industries and investors. Big data and the rollout of 5G are likely to mean the communications industry is responsible for around 20 percent electricity use by 2025. A shift in corporate focus among these firms is helping to drive the pivot toward renewables.

This issue is being embraced by the new cohort of investors.

CHAPTER 6

Market Research

Market research is about becoming aware of your competition, the market trends, and consumer buying patterns. This is very important for the entrepreneur, especially if you're trying to disrupt a market.

COVID-19: once-in-a-generation crisis and transformation. Have you started researching all this will do to your business? If yes, you are on the right track. If not, then you should as soon as possible when you finish reading this book.

- "When written in Chinese the word crisis is composed of two characters—one represents danger, the other represents opportunity."—John F Kennedy
- 'There are decades where nothing happens, and there are weeks when decades happen."—Vladimir Lenin
- "Never let a good crisis go to waste."—Winston Churchill

Let's start with the competition. Not only should you try to understand whose business you're trying to disrupt, but it's also important to learn from their mistakes. This is vital if you intend to take some of their market share. You don't want to make the same mistakes your competition has made, as that will set you back. If you want a piece of the pie, you've got to get to it quickly.

That's why understanding the competition is important for knowing how you should behave in the market. Sometimes, you play nice. Other times, you act like an animal. You become the beast and go after every client you can get.

COVID-19: Not Just a Temporary Crisis, a Permanent Disruptor

Historic global crises such as wars, revolutions, pandemics, and so on often feel like they put history on fast-forward. Processes that normally take decades or longer to play out unfold in a couple of weeks. Coronavirus is the political, economic, and psychological event of our lifetimes that will drive disruption and transformation for years to come. It will bring a radical transformation of the kind that occurs only once in a generation.

The crashes of the twentieth century—World War I/II, the Great Depression—were hard to understand at the time because their main causes were set in motion decades before. The backdrop to COVID-19 has been no different.

In the past few decades, globalization, air travel, and international supply chains have made human beings more physically connected than ever before. We believe a key legacy of coronavirus is that countries could seek greater self-sufficiency in their supply chains.

Every crisis leaves its innovative legacy: The Black Death, which killed as much as a third of Europe's population during the fourteenth century, led to severe labor scarcity. Resultantly higher wages eroded feudalism, encouraged innovation in agriculture technology and accumulation of capital, and ultimately spurred on the Industrial Revolution.

The Great Depression created a "waste not, want not" attitude that defined consumer patterns for decades. The 1970s oil supply shock led to the first efforts on energy conservation, efficiency, and independence. And more recently, the 2003 SARS epidemic helped jumpstart China's nascent e-commerce sector and ushered in the era of mobile payments in the country. The moral of the story is that humanity always innovates after crisis.

Take the automobile sector, for example. There are enough choices and different automobile brands around the world. But how much is the consumer willing to pay for an automobile? Are they interested in the name, the make, and the model? Or is the history of the company that they're intrigued by? Is it the fact that a certain car maker is a great manufacturer? Are they buying performance or utility? There are so many variables for the consumer to consider when buying a vehicle and the entrepreneur who wants to disrupt that market needs to understand all of those variables, which translates into understanding the competition.

By understanding all of these variables, you create the talking points you'll use to sell to your clients, to your investors, and to your own employees.

> "Every pandemic in history has changed the world. Pandemics are like rivers. They alter national boundaries. They interrupt national alliances. They restructure internal political systems. There's something about each one though that is unique, and it goes way beyond the number of deaths and casualties ..." Dr. Larry Brilliant, Chair of Skoll Foundation: Ending Pandemics Fund

Focusing on the competition could influence how you create your product or service because there are copycats who just mimic the competition. Their product may not be better; it could just be a duplicate that is offered at a different price point. The key is to make yourself unique.

I call this your DNA. Your framework. Your sole makeup. It's your unique blueprint, the soul of your company, which helps you assess the market so that you can understand if you're ahead of it; if you've got the timing right, and the economics involved.

Looking Beyond the Peak: a Legacy of Huge Data Creation

What will COVID-19 leave? Unlike other crises, COVID-19 will leave the legacy of massive data and knowledge creation in the medical, technological, and social fields, among others. While global knowledge typically doubles every two to three years, we can assume COVID-19 will be a major data accelerator, speeding up thematic disruption from decades away to only years away for certain themes.

If the economics are right, people are willing to pay a higher price for a product or service if it saves them money or time, makes them more productive, or makes their lives easier. Luxury products tend to be expensive. If the market is experiencing a downturn, it might not be the best timing for your product. So it's a matter of timing and understanding what your competition is doing to obtain clients.

Consumers used to pay with cash because it was the only way they could pay. Then businesses started letting them pay with checks or buy on credit.

Next, banks issued debit cards and people were allowed to buy products and services with those. In the 1970s, credit cards were introduced.

After that, mobile phones facilitated payments for many people. Now, some people are buying things with cryptocurrencies. It's the evolution of payment systems. Do bear in mind that every generation has a different approach to consumerism and your business must adapt to such changes.

Furthermore, we think one should not fear the expansion of big government post-coronavirus. For it was from the world wars that the modern-day social contract across health care, pensions, and public services emerged. Governments that were trying to restore economic and living conditions and fight off mass unemployment intervened more in the economy and accelerated their building of social insurance systems and public services in health, education, and housing. We believe coronavirus will spur a reimagination of the social contract that reduces inequality. And, finally, in many ways preventing pandemics is similar to preventing terrorism. We know roughly where it can originate and what's responsible, but we don't know exactly when the next incident will happen. Hence, the approach to tackling them can be similar—by improving preparation and funding of safety measures before the next crisis to protect civilians to the best of the government's abilities.

The monetary system continues to evolve. Money is circulated in the economy differently today than it was one hundred years ago. The entrepreneur must understand this because these payment systems are how you're going to receive payment for your products and services. That's where your revenue will come from. In other words, your business will generate revenue based on how well you can employ your payment systems.

Secondly, today's economy is very digitized. Assets have been heavily datafied. In that environment, you've got to think about security. You must deploy the best protections for your company, your data, and client information against cyberattacks. This ties in with the competition because to be competitive now, in virtually any market, you've got to look out for your client. That doesn't just mean providing a product or service they want. It also means protecting their privacy and data as well as your own.

Where To Find Competitive Information

The entrepreneur needs to know where to go to find information on the competition. Where do you go, how do you retrieve it, and how often do you search?

Every business sector has analysts around the world who watch the industry and report on what the players are doing. They provide monthly and quarterly reports on how companies are performing, the technological advances taking place, what clients the businesses target, the demographics of the marketplace, and more. Keeping up with this information in your industry is very important.

By understanding the dynamics in the market, the entrepreneur can more effectively target the right corner of the market and better communicate your market position to potential investors. Remember, you have to educate the investor on the market.

By the way, to set up your benchmark, you must find all filings within the U.S. Securities and Exchange Commission, as all publicly traded companies file quarterly and annual reports. Filings are easily found at sec.gov/edgar. Learn from those.

Entrepreneurs are not analysts. It's not what they do for a living. They aren't watching every segment of the market constantly. Their focus is on other things. But they do have to gather information on the market and that can be tricky.

The best approach is to find the information where the government regulators already have it. At times, I tell students all they need to do is read the government websites and understand the way firms are making money and stay transparent with investors.

You can't just perform a Google search, read an article on a major competitor in your market, and think that is verifiable information. You have to go deeper than that.

Whether you are researching the market for your business plan, an investor pitch deck, or for your own personal understanding of the market, you must ensure the data you use comes from a well-known source and highly respected source.

When you create your business plan, the data you use must be reliable information. Investors want to know that you are consulting experts in your industry. The same goes for your SWOT analysis. Only cite the best sources.

We are now truly witnessing a transforming world. We believe this crisis is a call to action for a new framework, which is more important and relevant than ever in the current climate. The megatrends mentioned in our Decade Ahead:

2020s report will happen at a more intense pace than originally predicted and offer important solutions for tackling the greatest challenge of our time.

We see themes further driving and disrupting asset markets in the 2020s. Looking forward, we believe that thematic investing can outperform in the brave new post-coronavirus world: people—on the scarcer allocation of human resources, innovation—on the faster disruptive role of technology, markets—on greater allocation of scarce financial capital, government—on a bigger role of state public policy, and finally Earth—for the cleaner allocation of natural resources.

You can collect information on private companies if you can find it; however, competitive data is more readily available for public companies. If they're listed on a stock exchange, then they've gone through all the layers of due diligence to get there.

Another thing to keep an eye on when performing research is merchant acquisitions. Any type of merchant acquisition helps you understand whether the market is ripe for growth. When small companies hit the market with the right product but don't have the financing behind them to expand into that market, they end up selling their companies to bigger ventures.

Mergers and acquisitions are two potentially good ways to understand the public market. It also lets you know what the private market is doing.

The private market sector is tricky because the information is not available to the public. For many of those companies, all you'll ever get from them is whatever you hear on the news or whatever social media you can find. It's the companies themselves talking, showing you what they've got. Or a PR agency.

The Great Acceleration: We believe coronavirus has turbocharged many of the thematic megatrends. The Great Moderation was the reduction in the volatility of business cycle fluctuations in developed countries that started in the mid-1980s, compared with the decades before, and arguably could be extrapolated to the 2010s post the global financial crisis, up until the onset of COVID-19.

Conversely, the Great Acceleration is the concept of accelerating change in tech. It refers to the dramatic continual and roughly simultaneous surge in growth rates across a large range of measures of human activity, first recorded in the mid-twentieth century and continuing to this day. This will be further accelerated by COVID-19, where we see a moon-shot technology revolution rather than evolution.

All of that is to say that you've got to be careful about what kind of data you read and how you use it. Information on private companies may not be reliable.

I've been in the banking system for over twenty years. I follow every well-known bank: Goldman Sachs, JPMorgan Chase, TD Bank, Bank of America, Merrill Lynch, and others' transactions in the paradigms. I have been introducing them chapter by chapter for the reader. Any of the top banks and reports from those banks will be closely correlated.

In other words, their expert analysts use technical and fundamental analysis to analyze the players in the market all day long, every day. They're following the market day in and day out.

The information from those banks should be as accurate as possible because there is often some crossover. If you see the same information from two or more analysts, that's verifiable data and you can also use other sources, such as media outlets, to verify the data.

Some publications, such as Bloomberg and Munifax, report on stocks and bonds. This will aid you in putting together an exit strategy.

Shifts in global power have been accelerated by COVID-19. The pandemic is likely to further accelerate global influence being shared across more countries, including China. Peak globalization, the splinternet tech war, and now the diverging responses of the US and China to COVID-19 have highlighted the growing split in governance opinions across the globe.

Furthermore, combined with the declining influence of OPEC owing to the continued rise of renewables, this could result in some of the largest shifts in power ever seen in modern economic history.

Many times, though, media outlets will write articles based on what some company's public relations team tells them to write about. The PR firm will send a press release and the media outlet just publishes the press release or reworks it into an article. That's not trustworthy information, so the entrepreneur has to be careful with information received from the media. It must be verified with reliable data from a respectable analyst with a good reputation.

You should get in the habit of collecting and analyzing information on the competition and your market. Read something every day and look for reports and other market data. Set aside time on your calendar to do this.

Many entrepreneurs go through this process when writing their business plan then never do it again; that's not good. You want to keep up with market changes and pivots that your competition makes. Always gather this kind of information; never stop doing it.

Competitive research and market research are an important part of the business plan and the SWOT analysis, but it doesn't end there. Everyone wants to see if you know what you're talking about. For that reason, you've got to stay on top of the market trends. You can't stop learning, ever.

Market Trends and Consumer Buying Patterns

When it comes to geopolitics and globalization, the virus has added fuel to the fire.

Did you know that China's global share of manufacturing value has risen from 3 percent in 1990 to 25 percent? This has fueled anti-globalization and geopolitical tensions.

Key takeaway: We're seeing a faster shift from China as supply chain fears compound concerns over inequality, technology, privacy, and climate change. But the biggest impact is likely to be rising geopolitical tensions between the US and China, the West versus the East. Shift to DM (automation, critical areas, jobs) or other EM (diversification, labor). Surveys tell us that 80 percent of analysts see geopolitics/tariffs as the key de-globalization driver (pandemics 53 percent, technology 24 percent). Over a third of analysts have seen or are expecting reshoring or diversification in supply chains, while half expect climate-friendly investments to increase post-COVID-19.

Market trends and consumer buying patterns occur on a generational level.

Every single generation in our society has a different way of consuming products and services. You have to understand who your future buyers are going to be. It's also important to understand past generations and how they used to buy things. When you place yourself in the market, unless you're targeting a specific generation, you should place yourself somewhere in between so you don't completely lose a generation. Today that means having something to offer for the Greatest Generation as well as Generation Z, and all the generations in between.

Generation Z is the future consumer. It's important to understand that. They are just beginning to come of age and are entering the buying market. Educate your investors on how they are buying and what they're buying.

Technology: Seizing the Opportunity, Enabling the Recovery

Video conferencing tool Zoom saw average daily users rise 30x from January to April 2020, and up to 60 percent of salary-paid US workers could work from home (US Bureau of Labor Statistics).

Key takeaway: COVID-19 has seen an acceleration of structural shifts in how we use and deploy technology, leaving a legacy of advanced computing, data generation, new infrastructure, and bandwidth. Commercializing moonshot technologies such as autonomous vehicles, quantum computing, and vertical farming could also be accelerated. Work from home, e-commerce, and stay-at-home activities are nearer-term shifts. Challenges include constrained demand, funding, and further erosion of privacy.

National surveys tell us that half of sector analysts expect technology investments in their sectors to increase, with 44 percent expecting more robots and automation.

This is easy to see when you're a parent. You'll see your children and how they interact with their peers. Whether it's communication, video gaming, the way they dress, how they perceive society, or what they think about the news. Those are all market trends and impact consumer buying patterns. It's important to identify that.

When I see my son wearing a certain kind of shoe, I ask, "Why is he buying that?" Is it because he's following the music industry and mimicking a musician he admires?

Recognize the patterns as they are happening on a private level. I mean, where you're at. Look at your children and the children of your peers. Observe yourself and your friends. Look at your parents and their siblings and friends. See how the different generations are interacting with each other. Look for the clues that can inform you on where your product can fit into the market.

I started a motorcycle business. Before I started the company, I asked, "Who's riding today?"

If you notice, Harley Davidson targets its motorcycles to men. They use women in bikinis and skimpy outfits to model for the motorcycles because they're targeting men. But the world has changed. Today, women are also riding motorcycles.

It's what I call the "sheconomy." When you visit the Mediterranean countries—Italy, France, Switzerland, Spain, Portugal, and Greece—women are riding scooters to get to work. In fact, there are more women riding motorcycles in Europe today than there are men. But when you look at the US market, it's mostly men riding motorcycles, and women are still portrayed as sex symbols. That signifies a gap between the genders, which ought to be recognized.

In our fashion brand, we'll be targeting pink dollars, that's the gay community. The gay community happens to be the second- largest spending consumer, right after women. Men are the last ones to spend money on fashion. There are many reasons for that—psychological, emotional, and generational, even taste and sophistication—but those are the markets. In fashion, these are the trends and the patterns that dictate the future of the fashion industry. As an entrepreneur, you've got to pay attention to that.

Following and analyzing market trends is important because an entrepreneur can have an amazing idea for a product, but if the market changes and you don't adapt, then you'll become obsolete. You can do a lot of work and invest a lot of money only to be relegated to history.

Blockbuster is a good example. It was hugely successful, but they didn't see Netflix coming in and changing the market. Now, the DVD market is smaller because Netflix came along and offered something better. The market changed and Blockbuster didn't see it coming.

My generation listened to cassette tapes. I used to sit down with a radio and an empty cassette tape and record a song so I could listen to it whenever I wanted to. I hated it when a commercial interrupted a recording because I had to pause the recording. I recorded several songs at once. But when my favorite song came on the radio, I'd hit record and I could listen to it any time I wanted to.

In the 1990s, CDs came around. I remember the first guy in my neighborhood to get one. He's walking around with a CD in his pocket like it was the best thing ever and people would ask, "What the hell is that?"

"Music," he said.

Next thing you know, it was trending. Everyone was getting one, and then these big music stores popped up and you could listen to the music right there in the store. When my generation came of age, CDs were the next best thing. Today, it's Spotify and iTunes. You buy the Mp3 or Mp4. You have a master player where you can download all the songs on the Internet you want at any time, and it's cheap, fast music. Do we not miss Napster?

Technology has shown us that the way we explore music in our culture has changed. Today, if you have Spotify in your car and your car has a smart player, you can listen to it streaming from the Internet. You can have a connected TV in your car.

Connectivity has become a huge thing. Look at telecommunications. Just like music, information is being relayed through these connected devices. People use nanosatellites. Why?

It's because of technological advancement. When Apollo 13 went to the moon, that technology is the same technology we have today in our mobile phones. What does that tell you? We don't need huge radios, big computers, or large technological gizmos anymore. The little cell phone we use today has the same computing capacity we used to send people to the moon.

To make that happen, our connections have to be faster, more durable, and more reliable. Those are the trends in the tech market. The faster information, music, and data get to us, the better it is. See how society changes our mindset? See how it changes the way we behave?

You can hang out with a friend of yours and they may listen to music on cassette tapes or LP disks. If you're old enough, you remember vinyl records. You think this is cool, but why would you do that if you can listen to Spotify on your phone? You can connect anywhere in your house. You have a smart TV, smart computer, smart devices in the bathroom, an iPhone, and everything else is smart.

These are the market signals. They're right in our faces. But we don't read them as trends or patterns until we analyze them. Now the question is, how do you disrupt that? What comes after smart everything?

The entrepreneur must think about these things and work outside the box. Try to come up with the next big, cool, fast, and reliable trend. The only way you can do that is to pay attention to the market. Put your ear to the ground and listen. What is the competition up to? What are the market trends and consumer buying patterns taking place right in front of your eyes? Get out in front of that.

Keep calm and carry on (contactless). While the economic disruption from COVID-19 was unexpected and the impacts felt across most industry sectors, technology has been critical in enabling businesses to continue operating, albeit with several needing to adapt. The crisis has highlighted the potential and pitfalls of certain technologies as alternatives to what we're used to.

Disruptive times require disruptive minds. In this section, we consider the technologies and services that have been prevalent for consumer or business use in the wake of COVID-19 and their potential long-term implications. We believe some of the behavioral changes exhibited in the crisis will accelerate structural shifts already underway before the crisis. Thus, while the short-term economic impact has been negative, there will be beneficiaries from these transitions, with regard to both existing companies and new deployment of technology to capitalize on these trends in the years ahead.

Paradigm Five: Demographics

The global population will increase by approximately 1 billion people and the demographic mix will have changed tremendously by 2030. We are reaching "peak youth" with a rapidly aging population globally, seeing Gen Z overtake Millennials as the largest generational cohort, and witnessing the emergence of the EM middle-class consumer in Asia.

Furthermore, the new consumer's preferences are shifting away from the old economy towards tech-compatibility, sustainability, and experiences.

World population: 1 billion in 1800 vs 8.5 billion by 2030. In the year 1000 AD, the world population was 256 million people. In 1999, 1,000 years later, it was 6 billion. Today, the world population is 7.6 billion and is expected to grow to 8.5 billion by 2030, and almost 10 billion by 2050. However, the two most important demographic trends for investors in the coming decade are: (1) a slowdown in the growth of the working population; and (2) rapid aging of the global population.

Peak youth: more grandparents than children. In the present decade, the world will experience "peak youth" for the first time in human history, with the number of persons aged 65+ expected to outnumber children under age 5. From 1950 to 2015, under-5s rose from 340 million to 670 million; those aged 65+ from 130 million to 610 million. By 2100, the UN expects the number of under-5s to remain flat at 650 million, but those aged 65+ to reach 2.5 billion.

Life expectancy increases: the "miracle" of our lifetime. Life expectancy is rising rapidly thanks to technology and better lifestyles, and has been one of the most remarkable demographic stories of the past century. The global average life expectancy has increased from age 31 in 1900 to age 72 today and is projected to reach age 83 by 2100. The period between 2000 and 2016 saw the fastest increase in life expectancy since the 1960s (source: UN, WHO).

Aging: enter the silver generation. In the next 10 days, 128,000 people will reach retirement age in the US, Europe, and Japan. By 2050 there will be 2.1 billion people over 65, and just four workers for every retiree versus eight today.

A key implication of an aging population is that an economy's savings rate tends to rise, while increases in investment levels tend to decline. The global pension savings gap is set to soar at 5 percent per year, from US$70 trillion in 2015 to US$400 trillion by 2050, driven by India, China, and the US. Japan's aging portends high savings and very low interest rates in the 2020s (according to OECD). Unlike Japan, Americans save via equities, making wealth management for "The Silver Generation" (600 million to 2.1 billion from 2020-50) a key financial markets trend. Overall, the aging of the population is encouraging saving rather than spending, which is deflationary rather than inflationary.

Future Human: immortal combat. "Immortality" may prove the most interesting secular theme in the 2020s. Life expectancy is rising rapidly thanks to technology and more. It is not just about living longer, but also about living healthier with a better quality of life. We are entering an era of "healthceleration" where advances in technology could bring about a quantum leap in the quality and length of human lives. Innovations in genomics, big data & AI, and "ammortality" mean that living healthily past age 100 could become more common in the coming decades.

The next frontier in precision medicine will further extend life expectancy and quality. For instance, aging has not traditionally been viewed as a disease that can be treated, but this trend is changing with more scientific breakthroughs.

Hence, longer-term, moon-shot medicine could result in more radical life extension that "disrupts death" by tackling the hallmark pathways of aging such as genomic instability, telomere attrition, cellular senescence among others.

Investment: Gen Z and the New Consumer (tech, sharing, experiences). We believe our demographics theme is the ultimate demand driver in a transforming world and will profoundly impact the consumer sectors over the next 10 years. However, consumers in each demographic cohort are as diverse as ever, meaning there is no silver-bullet strategy to fulfill demand. Companies will need to adapt or face disruption from the emerging spending power not just of Millennials but also of new Gen Z consumers (e-commerce, same-day delivery).

But there are also "untapped" opportunities. The aging consumers are the #1 cohort by spending power, but less than 5 percent of targeted ad dollars are directed to them. "Structural" opportunities are available via the "Bottom Billions Consumers," the 3.6 billion middle-class consumers that are expected to grow to 5.3 billion by 2030.

The key is KYC: Know Your Client.

Gen Z: Millennials Are No Longer the Largest Cohort

Generation	Greatest / Silent	Baby Boomers	Gen X	Millennials	Gen Z
Years Born	1923 - 1945	1946 - 1965	1965 - 1980	1980 - 2000	2000 - Present
Population (Global)	0.3 bn	1.1 bn	1.5 bn	2 bn	2.4 bn
% of Global Population	5%	15%	20%	27%	32%
Life-Defining Events	World War I and II Great Depression Electric Appliances	Cold War Moon Landing Transistor Invented	End of Cold War Live Aid First Personal Computer	9/11 Terrorist Attacks Global Financial Crisis Social Media	Post-Great Recession MeToo / LGBT Movement Electric / Driverless Cars
Communication Style	Letter	Telephone	Email/SMS	Instant Message	Emojis
Key Technology	Car	TV	PC	Smartphone	AR / VR
Hobby	Reading	Watching TV	Surfing the Internet	Music Streaming	eSports
Digital Proficiency	Pre-Digital	Digital Immigrants	Early Digital Adopters	Digital Natives	Digital Innates
Iconic Figure	Franklin D. Roosevelt	John F. Kennedy	Kurt Cobain	Mark Zuckerberg	Greta Thunberg
Music	Jazz Swing	Elvis The Beatles	Queen Madonna	Beyonce Coldplay	Billie Eilish Justin Bieber
How They Get Around	'55 Ford Thunderbird	SUV	Bicycle / Car	Uber / Lyft	Mom's Prius
Current Living Situation	Retirement Home	Semi Detached House	Own Small Apartment	Sharing an Apartment	Parents' House
Social Network other than Facebook	The Rotary Club	Match.com	LinkedIn	Instagram	TikTok
Deepest Fear	The World Today	No longer center of attention	What about my generation	Paying off Student Debt	Low Batteries
Key Life Question	How did the country go so wrong?	What will I do when I retire?	What's the point?	What's a career?	What's a landline?
Defining Condition	PTSD	Dementia	Stress	Gluten-Intolerant	Peanut Allergy
What They Spend On	Oklahoma Community Dinner Theater	VIP Tickets to The Rolling Stones	Radiohead Concert	Coachella	Fortnite

CHAPTER 7

Branding, Marketing, and Advertising

Many entrepreneurs use the terms **branding, marketing, and advertising** interchangeably. They talk as if they are the same thing. There are nuances, but each term references a different aspect of creating the business and the perception of your business.

It's important to understand the meaning of these terms if you're going to use them. You're building a business, after all. If your goal is to disrupt a market, then you need to understand the different terms and how to deploy the concepts. So, what do branding, marketing, and advertising actually mean today? What are we talking about when we use these terms?

Branding is the overall picture of how your name, your brand, and your company are positioned in the market. Of course, there are multiple ways to position yourself in the market.

You can be a high-end luxury company. That's one way.

Another way is to position yourself as a new international global platform for other brands to come to.

You can be a fashion brand that everybody likes to have access to.

Marketing is about how you're communicating to the world, to your client, and to your investors. It's about how you communicate who you are. It has to do with communicating what your products and services are, why you're the best company to offer them, and why you are in the market segment that you chose for your company. I will use wine as an example.

Oversaturated with the number of wine brands in the world, especially in the US, a quick reference as of 2021, the US produces 3 billion bottles of wine and consumes 4.7 billion bottles of wine annually. Yes, we all have our favorite wine, our "go-to" brand. But when you see a wine list and want to try something new, you pick the one you want to try. You might say, "I've never had or heard of that one. Let's try it." But, why?

Perhaps you heard from a friend that it was a good wine or perhaps the waiter did a good job selling you on the brand, so you try it. The communication part of that is marketing.

You're willing to try it out because someone communicated to you the benefits of drinking that particular wine. If it really is a good wine, you'll order it again and again, maybe even recommend it. But guess what? You've converted. That is what we call in the sector a client acquisition strategy.

The communication part of marketing is converting a client from an existing brand, maybe even loyalty to that brand, to your brand. That's not easy, but companies do it all the time. What has changed today is that Generation Z members are not loyalists anymore and I will explain more to you what that means.

Advertising is like traditional commercials—short clips of a video, short messages, or communications where you intend to drive traffic to your website or physical store.

The website is the storefront for commercial businesses today. If you don't have a website, you don't exist. People are more likely to visit your website before they visit your physical location. That's important to remember when advertising your brand. The fact that you can search on various platforms to find a location then read someone else's reviews or testimonials to decide.

I recall as a child that taking a vacation with my parents always involved the same places they once knew or had been to. I always wondered why we didn't ever go somewhere else in those towns. It was based on the relationship my parents had with the owners of those places or their own experiences that needed to be shared with us.

About twenty years ago, in the early nineties, I met this girl in marketing with a strong Californian accent pitching me some ideas. She came to me and said, "Marco, I couldn't find you on Google."

I laughed and said, "Good! What the hell is that?"

She said, "You're not Googleable, and that means you don't exist."

Well, I existed. I was standing right in front of her. But her point was, the world has changed. That was twenty years ago. Twenty years before that, there was no commercial nor Internet and nobody had websites. Advertising was different. Driving traffic meant driving traffic to your physical business location.

When you see those guys in your town handing out pamphlets about their business or wearing a sign about it, those are dead business practices. Today, you look at a business website, visit its LinkedIn profile, and you evaluate its achievements. All of that is advertising. It says who you are as a founder, an entrepreneur, and a company so that people can get to know you.

People will get to know you through those channels; then they'll start reading more about you. The more they read about you, the more convinced they are that you have a great company, a great product, and that they should buy your product or service. Maybe they'll even convert from their brand loyalty to yours next.

Don't get hung up on which comes first. Chicken or egg? It depends on your budget and your values.

Today, you can do all three very inexpensively using social media and other digital channels. There's Instagram, Facebook, LinkedIn, and all the other social and digital channels similar to those, but you have to be precise in understanding your target market. What age group are you targeting? What gender? What generation? Figure out which social media channels your target market is using and join them.

If you use those social media channels and your target market is younger people, you're not targeting the right client in the right way.

If you have extra money in the budget, you can put some toward social media algorithms driven by artificial intelligence. AI specifically targets whoever is clicking or searching for a product, or involved in the sector that you're in. You can use that technology to drive traffic to your website, get people interested in your brand, and convert them with the right marketing tools while positioning your brand in the market simultaneously. It doesn't matter whether you're targeting Europe, Asia, the US, or going global. It works.

The more money you have, the more effective you can be at positioning yourself. That said, if you have a product that you want to prove to your community, localize it. If you can convince your local folks to use your product, then you have a viable product. You can expand from that base.

You have a base of clientele. Start there because that's where you can learn, make mistakes, and adjust. Another thing that I would say is that there is no one strategy that is the right strategy. You have to be flexible, and you have to understand that you do not have all the answers to every problem. It's an ongoing journey, the journey of an idea. Be flexible, mold it into what the market is asking for, and then you can go in for the kill.

The Importance of Brand, Product, and Service Names

Your name has to speak volumes to the customer. Whether it's the name of your product, the name of your company, or the name of a service, it has to speak on multiple levels.

For example, what is so great about Italian products is that they use all five senses to ideate a brand name. Your brand name portrays your product. You want it to represent the product in the right way. One brand that has done that well is Vespa.

Vespa is a scooter, but the word means "bee." It's a type of hornet. The Vespa's engine sounds just like the flying insect it is named after. When you see the name, it triggers in your brain the image of a bee. It's that connection of the word in your brain that causes you to think of the brand. You think of a scooter. The branding is subliminal in all aspects. That illustrates how important it is.

Another way to determine if your branding is working is to think about your legacy. We all have surnames and a history. We like to be recognized, and we want to leave something behind.

We tend to think of ourselves, with our egotistical mindsets, as number one. When you're promoting a brand with your name on it, that can be great. You can push the product; then you're going to have to push yourself more than you would normally. If the brand was someone else's name, you might not be as excited about it. But, because it's your name, your reputation, and your legacy on the line, you want to make sure that you leave behind something that people will remember you by.

In the case of Soriano Motori, the word Soriano is a breed of cat. That became the nickname to all family members. We used that to maintain feline names for all models we created; even my son is called a cat.

All the big names in the US will verify this. When you think about the Carnegies, the Vanderbilts, the Colgates, the Rockefellers, and even the Kennedys, they all left behind their surnames as a brand name to represent their companies and historical footprint. They were careful in doing what they did, but some of them were monopolistic. Highly illegal to the point that it created antitrust laws in our constitution. It did not matter.

Even after several decades of industrial growth in the United States, these families are still looked at as the best families for several generations. Most people refer to them as the doers.

Fashion is one industry dominated by well-known names. You recognize Versace, Armani, Michael Kors, Calvin Klein, Hugo Boss, and Tommy Hilfiger. We recognize them because the individuals who created the companies were market disruptors with their trends. Their ego plus an entrepreneurial mind created wonder.

If you knew that it was Hugo Boss who was commissioned by the German Nazi military to make their uniforms, would that change your mind to buy the brand? I've seen several Jewish friends of mine buy the brand while they complained about the war atrocities. We are surrounded by these contrasts in life of our philosophies versus what we choose to consume even today.

To think that you're number one helps, but you have to make sure that you do it right. If you leave behind a legacy, especially with clothing, and influence what people wear, your customers will become ambassadors for your brand. You want to make sure that you leave behind something amazing. We are ultimately a conduit for others to talk about.

The Soriano firm does that with our fashion brand and also the motor company we've rebirthed. Our goal is to make these brands number one, so that no matter where you look, you see them, you hear them, you read about them, you smell them, and you feel them. Recognition of the name will remind you that Soriano is a strong brand.

The name Soriano is a breed of cat, a tabby cat. All of the products that we've manufactured in the past, and that we're manufacturing today, have a feline name. Whether a panther, a tiger, lynx, or a jaguar, we're sticking with the cat motif. Who doesn't love cats?

I think of myself as partially humble, but I'm also partly crazy because you have to be a little mad to be an entrepreneur. You have to be bold and obsessive. You have to think you are the only family member who can do it. And, by the way, that means you work twenty-four hours, seven days, 365 days a year.

The Soriano family has numerous members, but I was the trailblazer. When I was about seven years old, we'd go to the beach and I'd be the first one out of the car. I'd walk a mile down the beachfront and other members of my family would wonder where in the heavens I was going.

I was always independent, ready to be on my own. If I wanted to go to the beach, I'd just go all by myself before everyone else. Today, if a child walks more than ten feet away from their parents, they will freak out. Back then, Mom didn't freak out. I was gone, partially humble, and partially crazy. I wasn't afraid to be a trailblazer and Mom knew what she had given birth to.

Keep in mind that there are ways you can hurt your brand, too. Politics is one place where a brand can be destroyed. One that comes to mind is our former president, Donald Trump. He was well known for having great hotels and being a mogul of the industry. It was a well-recognized brand name, especially in New York City, before he got involved in politics.

Entering politics segregated him from a significant slice of the population and it hurt his brand. Yet, thousands of people would want a picture with him or have their picture taken in his buildings.

Entrepreneurial behavior must be watched. We are leaders in the industry, and authorities as well. That means we must behave responsibly because a lot of other people will look up to us. We become role models and that implies accountability. They'll watch what we do, listen to what we say, and witness how we behave. They sometimes even mimic our actions.

When you play the influence card, you have to be aware of how people perceive you. Being socially responsible with your employees, your brand, and your client is very important. Today, someone with over five thousand followers on any social media must be extremely careful with ideologies. We are all connected regardless of our perception.

The entrepreneur must also realize it's not all about him or her. It's not just one generation you have to worry about. Play the long game. In other words, breed future generations into the game plan. One of the humblest moments I get is when the spouses or children of our employees come to me with such love and thank me for what we do. They are grateful.

At family gatherings, for instance, I'll take time out to be with our children. I'll take them aside for an hour and discuss whatever family business I'm involved in with them. I like to ask them questions and let them ask me questions, as well, so I can have a clear understanding of how they perceive the family business. Do they have a growing interest in carrying the flag? I record the meeting and archive it for them to see one day and understand what these initiatives meant.

Not every child does. They all have different personalities. But I try to recognize which ones have it in them, which ones are taking an interest, and which ones might potentially replace me. In other words, who will be my successor? Which child in the family will embrace the brand and the family business and help it grow?

We have nine children in my family's fifth generation right now. For sure, I have identified two of them.

Taking Your Brand to the Level of Influence

I call the Soriano brand a brand of brands. When we established our platform, we did it in a way that allowed us to take any mode of transportation off the road and power it with our unique invention of the modular EV system.

That's a huge statement to make. To make it happen, I ensured that our investment was put into building out an industrial, global platform. I didn't just want to be a manufacturer. I didn't want to compete with the big brands of the world. I wanted to serve a high-end niche demand. I also wanted it to be Italian by design, engineering, statics, branding, advertising, marketing, everything. That was done on purpose.

Number one, the trend in Italian products is that they tend to be high-end luxury items. Italian products have the best reputation in the world. Look at vehicles, for instance. When you think about high-end luxury vehicles—super sports cars, motorcycles, and boats—the ones at the top of the line are Italian. Then, after that, German. Third place, the US, maybe.

I believe we can help many of those brands reposition themselves with a futuristic view. I'm being partially humble in saying we have done a phenomenal job. We're ahead of our time in some of the designs that we've come up with.

Harley Davidson, for example, is one of my favorite motorcycle brands. For the last ten years, management has been destroying the brand. But I think we can save it and maybe even own it, or a piece of it.

This is the partially crazy part of me. I would love to have a brand like that and rebuild it the way it is supposed to be.

Brands need to understand that, in order to have the DNA, the blueprint, they can't just think about reducing costs by outsourcing to emerging markets; that takes away quality control. It takes away so much of the DNA, what made those brands what they are in the first place, and I can proudly say that Soriano Motori is 100 percent Italian.

Here is a part that I do not support locally as Asian brands have saturated our markets, but we local producers are limited to what we can sell. These are the laws that impede local manufacturing from being more successful and sustainable. For example, Honda is allowed to sell over ten thousand motorbikes per month in Milan, and we are only allowed to sell four hundred.

Branding, and how good you are at it, depends a lot on which side of the brain you use. Are you highly creative, right-brained, or totally logical and left-brained? Unfortunately, not everybody has a good balance of both sides of the brain. But it is remediable.

Part of the training that I had growing up was artistic. I had a very progressive education, which exposed me to the arts, music, and all subject matters. Even mathematics and science. Chemistry. Also, as I mentioned, I speak multiple languages.

To breed the next generation of entrepreneurs, or to expand your legacy, you need to focus on education first. You want to expand the brain, cultivate their neurons in the future generations of thinkers and doers. Start with yourself. It's not too late. Then you can pass it on to the next generation, who are the future brains, future leaders, and future thinkers of the world. You want to breed those qualities for the future. To do that, you've got to have a level of influence.

If you have a great company, you'll attract great minds to it. That's going to make you grow. Be aware of that potential when you're branding yourself.

Some people think branding is a waste of time. It's about showing off, showboating, or whatever you want to call it. That's the wrong attitude because you're targeting a consumer demographic. You want a company that has influence in the market. You want your product or service to influence the customer beyond the cash register. The only way you can do that is to combine branding, marketing, and advertising into a comprehensive company-building mindset.

Don't Drink the Kool-Aid: You Are Not the Smartest One in the Room

A statistical error is a mistake in calculating a measurement or a sampling of a cohort and makes it difficult to draw a valid conclusion. Here's an example.

Let's say you're sitting in a room with ten other people, and you ask them all the same question. You're likely to get ten different answers. Let's say you ask them all to guess your age. If you take the ten answers and average them, that will come fairly close to the real answer. On the other hand, if you were in a room with one hundred other people and asked the same question, there would be less statistical error. In other words, a greater number of people would guess closer to your real age.

The larger the pool, the greater the connection between their brains' wavelengths.

In other words, the more people we involve resolving a problem, the better solution we can find.

As an entrepreneur, you want to stay connected to other people, other minds. You want to consider all ideas. Don't become isolated in your thinking, ever.

This is one of the reasons a company like Unilever exists; they put together focus groups. All the companies selling food and beverages have to consult focus groups.

Consider cereals, for instance. How many cereal varieties are there in the United States? The reason we have so many cereals is that cereal companies have focus groups and based on that, they know that children choose a particular type of cereal over another. Put aside the fact that many of them are sugar-based. But they exist because they've asked which flavors children like, and they make cereals based on that.

Focus groups show whether your idea has potential because you can put it in front of a large pool of potential consumers, give them a survey, and it will tell you if you have a market. If you understand how that works, you can avoid or minimize a statistical error by not thinking that your idea is the best idea in the world. The market will tell you if it's good or not. At this point, you can use your own social media to obtain free market opinion.

I am now in my forties, and I've managed to position myself as an expert, a thought leader, a teacher, a professor, a tutor, and a mentor to others. I can't speak for younger generations about how they perceive the world from their consumers' eyes, but I can talk to them and learn.

Teaching is not just about teaching. It's also about learning by talking to younger generations, or older, and listening to what they have to say. This is why education speaks very highly to me. In fact, it's so important that I want to create a foundation to help students obtain full educational scholarships and connect the dots between design, engineering, and fashion.

There's an element between these three industries called "beauty." If you don't understand what beauty or elegance means, then you don't know what you're doing. That's why a lot of companies fail. So, I want to help students understand that concept. After all, it's the future of the Soriano company and the future of generations to come. It is, hence, my legacy.

Branding is about positioning yourself. Speak out, have an opinion based on verifiable data and not just something you pull out of your ass. If I could charge people for the air they breathe before they start talking, I'd be a zillionaire. Useless talk doesn't solve problems and it won't build businesses. Data does. Base your reputation on what counts and brand yourself using information people can rely on.

Paradigm Six: Robots and Automation

Up to 50 percent of jobs are at risk of automation by 2035. Winners: automation, local production, big data, and AI; losers: humans, global supply chains.

Up to eight hundred million, or 50 percent, of world jobs could be at risk of replacement through automation by 2030–2035. The cost of industrial robots fell 27 percent from 2005 to 2014 and is expected to decrease around a further 22 percent by approximately 2025. On the flip side, the industrial robot market size is rising from US$38 billion in 2016 to an estimated US$71 billion by 2023, driven by robotic process automation (RPA), speed factories, and other use cases. We also enter the 2020s with underlying trends in big data and AI accelerating toward the exponential threshold of "technological singularity."

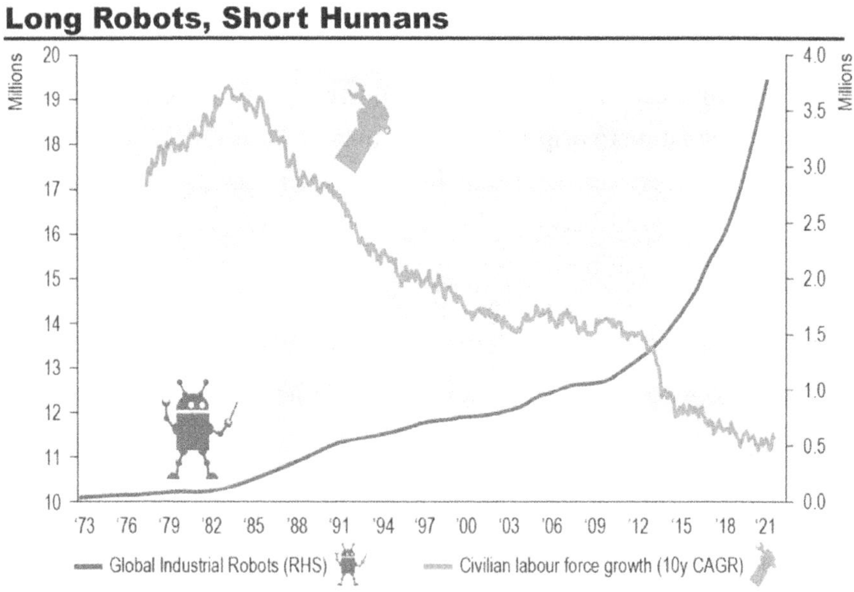

Source: BofAML Global Research, IFR, US BLS

Robotics and automation are not new concepts and have been used in various applications for decades now, for example, in manufacturing plants or in the office via software. But the next generation of robotics and automation could be even more transformative, putting up to 50 percent of jobs at risk by 2035 as well as transforming supply chains and economies. Enhanced computing and artificial intelligence are driving improvements in many diverse areas such as optical sensing, machine vision, voice recognition, environmental sensors, motion actuators, and touch/haptics that are set to take robots and automation to the next level, opening up new possibilities and job risks.

Furthermore, demographic issues such as aging and restrictions on immigration as well as trade tensions (trade wars, tariffs) also support a move toward more robots and automation.

The key trends in robotics and automation to look out for in the 2020s:

- Reshoring: possibility of reshoring with automated factories would reshape trade.
- Autonomous vehicles (AV): could bring new business opportunity, but also job risk.
- Artificial intelligence: AI solutions could impact a wide range of service jobs.
- Agricultural automation: precision agriculture technologies could impact chemical use and farm jobs.

Industrial robots and speed factories to change globalization: Industrial robotics is advancing fast and, while fully automated factories are not always possible in every type of manufacturing, an increasing number of activities can be automated, or products reconfigured for automated manufacturing. This trend diminishes the importance of manual labor and makes processes much more efficient. This shift could materially impact global supply chains, especially compared with the past few decades when cheap emerging market labor was the biggest driver of globalization. Now, advanced industrial automation can enable local production, unintentionally playing into rising nationalism and trade protectionism themes but also helping to reduce environmental footprint.

Traditional Supply Chain vs. SpeedFactory

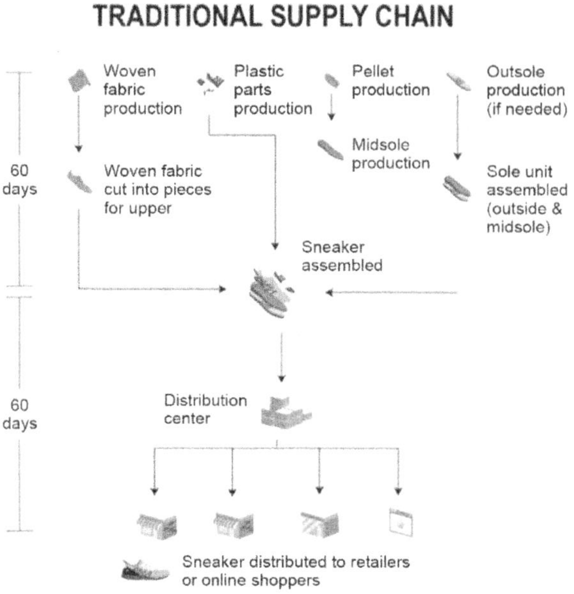

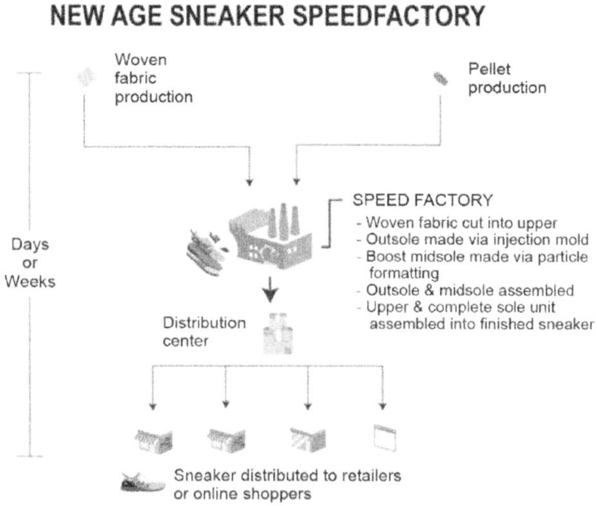

Source: BofAML Global Research

AI, Robotic Process Automation (RPA): Service jobs are at risk. While robotics and automation are typically associated with factories and industrial establishments, increasingly, office and service sector tasks are being automated due to improvements in computing and software. From voice text recognition to automated risk assessment, investing and medical imaging, many previously "human" tasks are being replaced by powerful software, some of which incorporate AI concepts like machine and deep learning, natural language processing, machine vision, and so on. Chatbots and voice assistants are well-known examples but, beyond these, new automation companies are deploying RPA or robotic process automation, which aims to use software to replace a whole host of human activities in the office (e.g., automating the accounts payable process). RPA companies are growing fast from a low base and may not only put office jobs at risk but also change the dynamics of outsourcing versus reshoring. But, by freeing up time from routine tasks, RPA could also enhance the value of remaining jobs.

Investment: robots and automation sector opportunities: the number of industrial robots in use worldwide is set to grow from 2.25 million today to 20 million by 2030 (Oxford Economics).

- The rise of artificial intelligence (AI): companies involved in AI and related process areas (RPA, natural language processing, enterprise software) could create significant value in the coming decade. The market size for AI analytics and technology could exceed an estimated US$110 billion by 2024 (source: BCG).
- Industrial robots are helped by lower costs and better tech. As factories modernize in emerging markets and wages rise (China 9x wage growth since 2000), the global industrial robot market is expected to rise from US$11 billion in 2014 to an estimated US$24 billion by 2025 (IFR, BCG).

- Autonomous vehicles opportunity with several auto companies, mobility providers, and tech firms vying to develop this technology and new AV business models.
- Aerospace and defense: unmanned automated vehicles (UAVs, drones) expenditure rose 10x 2000–2016 to US$3 billion and military robots (10 percent CAGR, US$14 billion market by 2020) are the opportunities due to safety, flexibility, and cost benefits (source: BCG).
- Financial services are deploying robots and AI, for example, robo-advisors, analytics, automated trading, fraud detection, and credit risk checking.
- Health care: deploying medical robots (accounting for 35 percent of professional services robots) and computer-assisted surgery. Telehealth and algorithms, care-bots, and bionics all offer significant robot and AI opportunities.
- Service robots perform services useful to the well-being of humans, assisting with jobs or chores. Growth will be driven by convenience and an aging population.
- Agriculture and mining are also seeing automation due to labor shortages and large-scale operations suitable for automation, e.g., driverless tractors/trucks, milkbots.

CHAPTER 8

Sustainability

Sustainability should be understood as a broad discipline. It should give investors, students, and entrepreneurs insight into business, technology, the environment, and other social sciences. The core skills that students are learning today in school are highly sought after, especially in the modern economy where many people around the world are trying to reduce carbon emissions and working on the development of future technologies. Sustainability draws from politics, economics, philosophy, and, of course, the social sciences.

The Outcome: Big(ger) Data, Better Infrastructure, Smart(er) Cities

The longer-lasting technological outcomes post-COVID-19 will derive from a combination of data and technology infrastructure, in our view. The structural shifts in retail and flexible working will lead to accelerated investment cycles in the enablers mentioned above: data storage, utilization of more computer and semiconductor capabilities, communication infrastructure (especially 5G), and more precise tracking, monitoring, and surveillance. We see three key implications of this:

- More robust communications backbone: increasing bandwidth deployment of 5G can accelerate new business models (e.g., OTT streaming with AR/VR, 5G-based cloud gaming, B2B services such as smart factories/hospitals) and the use of AI to handle this increased data.
- Big data and AI: creation, storage, analytics, and utilization of data at an accelerating pace, and enabling more disruptive innovation (accelerating the commercialization of moon shots such as autonomous vehicles, quantum computing).
- Smart cities: advanced potential for cities to use this new infrastructure and data for surveillance, mobility, safety, energy, and diagnostics. The vast majority of our sector analysts (75 percent) believe that big data and AI will be the most important technology for their sector post-coronavirus, followed by cloud computing (48 percent) and 5G (27 percent).

Sustainability skills—and that includes skills, talent, and environmental awareness—must be a priority in corporations and governments because it impacts the ability of future generations to meet their basic human needs. Fields that have a high impact on sustainability include civic planning, environmental consulting, building and construction, agriculture, and anything involving the natural environment. Corporate strategies, health assessment and planning, and the law are all important to sustainability.

There are three pillars to sustainability: economic development, social development, and environmental protection.

The Three Pillars of Sustainability

When Soriano Motori created a global EV platform, one of our main goals was to reduce greenhouse gas emissions because they cause climate change. That's undeniable because we've seen the impact everywhere.

Entrepreneurs must define sustainability within their businesses by outline objectives. There are about six different goals to keep in mind.

The first, a really big one, is to end poverty. Put an end to hunger. Secondly, create better standards of education and health care, particularly pertaining to water quality and sanitation. Third, gender equality. Fourth, creating sustainable economic growth while promoting jobs and stronger economies. Another goal is tackling the causes and effects of climate change, pollution, and other environmental hazards that harm people's health, livelihood, and lives. Finally, sustainability goals include the health of land, air, and sea.

Entrepreneurs who address these questions in their vision and mission statements only make their businesses stronger. Clients will want to grow with you. It's also attractive to investors to see that your business is about more than just making money.

A business plan should align your risks and rewards with common human interests, such as taking care of the environment. The enablers: techceleration in computing, data, AI, 5G, cybersecurity, and more. These ongoing structural shifts present several opportunities to the technologies that enable them. Indeed, despite economic headwinds, 50 percent of our sector analysts believe technology budgets for their sectors will increase post the crisis.

Economic Development

Economic development proves to be the most problematic of the pillars because there is such a wide array of disagreement in political ideologies regarding what is or is not economically sound, how that will affect businesses, and by extension, the impact on jobs. The economic development pillar provides incentives for certain types of businesses and other organizations to adhere to the sustainability guidelines beyond the normal legislative requirements. It encourages the average person to their little bit to achieve common societal goals. One person can rarely do much on their own but as a part of a group, that one person can do much more.

Data centers are the new real estate: Another lasting legacy could be further acceleration of data volume and processing requirements; the EU believes data volume will rise 5x between 2018 and 2025. Data centers provide the critical infrastructure that supports remote working, education, and TV streaming.

Additional capacity and connectivity are predicted longer term to support this accelerating structural shift, benefiting data center operators controlling network access points (to lower latency and improve connectivity), and wholesale providers through increased hyper-scale cloud capabilities.

Of course, economic development policies affect other areas of a community. Supply and demand, consumer behavior, and modern life, in general, require a lot of daily resources. For the sake of the environment, getting our consumption under control is paramount.

Economic development is about giving people what they want without compromising quality of life, especially in the developing world, while reducing the financial burden and bureaucratic red tape involved in doing the right thing.

Home Network Upgrades: +US$1.5 Billion

Despite home network broadband market revenues being broadly flat in 2017–2019, the market is projected to rise from US$3.5 billion to US$5 billion in 2019–2023 due to increased working from home, streaming, and demands on the network requiring better wireless broadband in all areas of the home.

Semiconductors are beneficiaries at all levels (consumption, computing, cloud). Our technology and semiconductor analysts highlight most of the technology-related enablers post-COVID benefit semiconductors. Remote working and e-commerce/home consumption could drive a corresponding increase in processing and bandwidth requirements, benefiting compute-levered semiconductor providers. Working from home has highlighted the need for enterprise resiliency and is speeding the transition to the thematic investing cloud. The need for better diagnostics, faster screening, and population monitoring is likely to increase artificial intelligence investments in health care and smart cities. Increasing online gaming will require better gaming components.

Furthermore, ongoing trade tensions are pushing China and the US to a path of greater self-sufficiency and supply-side duplication. All of which creates more demand for semiconductor equipment makers.

Cloud computing and cybersecurity become crucial: COVID-19 could accelerate the adoption of cloud computing as companies prioritize enterprise resiliency, operational redundancy, and resource flexibility.

On cybersecurity, today's technology teams believe that while virtual private networks (VPNs) have been extensively used for home working during the pandemic, 60 percent of companies will phase them out by 2023 owing to their inefficiencies, replaced by cloud-based connectivity and security software via zero-trust network access (ZTNA), which will be more cost-efficient and enable faster connection speed and data traffic closer to the user.

5G/Wi-Fi 6 and splinternet: China racing ahead on 5G? Greater home use is boosting demand for online gaming and pushing telecoms to move ahead on 5G and Wi-Fi 6 to improve the customer experience. But the bigger picture (and potential opportunity) post-COVID is the ongoing US-China tensions leading to potentially separate 5G ecosystems emerging, with considerable investment. Telecom teams believe China is racing ahead on 5G, set to roll out >630k 5G base stations by the end of 2020 to achieve "dominance in a digital era." These tensions and national security concerns over 5G infrastructure see an increasing risk of two 5G ecosystems emerging (supply chain, software), as governments are increasingly involved in selecting 5G vendors. This may delay the process of 5G rollout, with US/European applications and available capital expenditures limited in the short term.

Robotics and automation: Increased factory automation and logistics fulfillment could be accelerated in light of the crisis. Our survey showed 39 percent of analysts believed their companies would invest in more robotics and automation, in response to keeping operations open without humans. The long-term effects this will have on employment, therefore, could be negative.

Social Development

There are many facets to social development because it happens in different forms. First, awareness of it. Secondly, legislation can be used to protect people's health or mitigate health risks caused by pollution and other harmful business activities. For example, the developed world has strong checks and balances in place to ensure public health is protected.

Social development is also about maintaining access to basic resources without compromising quality of life.

Sustainable housing is a hot topic in this category. How can we better build the homes we live in from sustainable materials?

The final element of social development is education. I am a strong advocate for education, and it's one of the reasons I am creating the Marco Antonio Soriano Foundation. How do we encourage people to participate in environmental sustainability? How do we teach them about the effects of environmental protection while warning of the dangers of not achieving those goals?

Environmental Protection

Environmental protection is one of the most important pillars because the primary concern is the future of humanity. It defines how we study the environment, protect it, and sustain our resources. It impacts the integrity of our air quality and the quality of other natural resources such as water. The focus is on the elements that place undue stress on the environment.

We all know what we need to do to protect the environment. It's not a hard thing to grasp. Whether recycling or reducing our power consumption by turning our devices off instead of putting them on standby or by walking short distances instead of taking the bus. Regulations prevent business pollution by encouraging them to keep their carbon emissions low. These are the types of incentives we want to install in our homes and businesses for the purpose of renewing our power sources.

Lastly, it's about how technology drives our future. Today, the Environmental Protection Agency (EPA) recognizes that developing technology, and biotech, are key to sustainability. They have the potential to protect the environment from the negative consequences that technological advancements may bring.

How the 2020 Pandemic Changed Business

We live in unprecedented times with quarantines, social distancing, face masks, business closures, and rampant unemployment. People are stressed, anxious, and socially isolated. All of that impacts well-being. For some people, it develops resilience.

Resilience really is a good thing. It's the capacity to recover, respond, and bounce back from challenges. As a child, when you fell, your parents probably told you to get back up and get going again. Get going. Keep going. That says that resilience is a huge attribute of the human race. It's a built-in trait, embedded in our DNA.

Entrepreneurs must develop resilience. I call it the "bounce-back strategy."

First, understand that you have an instinct to connect. You need to get in touch with your feelings, acknowledge them. The pandemic has created challenges for a lot of people. We're humans and we're wired for social connection, so distancing ourselves from others is extremely tough. Being forced to do so was challenging for a lot of people, especially older folks.

I read a paper published by a German university in Munich. It explained the biology of threat perception in humans. Our natural response is to come together and seek safety in groups and communities, but the pandemic was the first time we faced a threat where we couldn't do that. We needed to stay apart, going against our natural instincts. That created discomfort, even dissonance.

Another strategy is to socialize as often as you can, even virtually. Sure, it's annoying. But it's important.

I read another study that concluded that communication reduces stress. Socializing with others reduces negative emotions. Whether you have a happy hour via video conference or you experience quality time across the street at the neighbor's house, socializing is a stress reliever.

Today, through technology, we have several ways to overcome isolation and create new ways to connect with each other. Our companies help us connect with other people daily, in multiple layers, through digital social media, virtual conferencing, phone calls, and texting. These tools have made a lot of difference for a lot of people. I forced my employees to engage each other through these channels because everyone was in isolation. It was important to do that.

A third strategy is to engage in new ways. If you can't be with others, try binge-watching your favorite shows, listening to your favorite music, or playing classic games. Doing activities that you're familiar with can give you solace, and it reinforces the feeling of being with old friends. For instance, when we listen to the same music over and over again, it brings us back to that one moment of happiness we remember when we were with a certain friend. It might be an old love, a memory of our parents, or something along those lines. That's healthy.

Stories also connect us. Even if you are already socially connected, your brain needs a story. That's why writing a book, writing a chapter, or writing something is important. It means you're paying attention to stories rather than statistics. You'll feel more engaged with others, too.

While listening to facts and figures, we use parts of our brain to form a new language of comprehension, but when we hear stories, we're engaging all seven parts of our brains, from the motor complex to sensory perception. That allows further ingenuity and leads to triumph in the mind.

The fifth bounce-back strategy is generosity. Be altruistic, give to or help others.

A part of being connected to a community is that you can provide for those in need, especially if you're in a position where you can. If you're not an altruistic kind of person, maybe this is an opportunity for you to become one. I've seen people grow into altruism during this challenging time. People were more helpful to each other, even if they were distant.

There was an event twenty years ago that provided an even greater sense of meaning and satisfaction to people's lives. It's called 9/11.

On September 11, 2001, a band of terrorists flew airplanes into the Twin Towers in New York City. The whole world rallied around New York at that time. Among the top behaviors people exhibited during that time was altruism. Restaurants gave food away and charities and family foundations tried to help other families affected by the attack. That's what a crisis does; it helps us grow to be more altruistic and generous.

Another important part of that to discuss is emotions. They have an effect on our interpretation of life. Essentially, we learn how to manage our perceptions. For example, people say, "I'll believe it when I see it." For whatever reason, we learn that seeing, touching, and feeling are what make something believable. That's not exactly the best way to manage our perceptions because we can also set ourselves, or others, up for disappointment.

I know that when I feel more positive, that's when I interpret information more positively. The pandemic, however, has its own implications. How do you keep your spirits up, for instance, if you have to keep reminding yourself that things aren't as bad as you think because the media is planting fear in everyone's minds? So you don't watch.

It's possible that, with our own anxieties, we can interpret things in a negative light. We might make bad decisions that negatively impact our lives. The bottom line is, stay positive. If you do, things around you will feel brighter.

Here's another bounce-back strategy: Pay attention to the basics. This is a good one because stress leads to more stress. Take a break. Sometimes, a quarantine causes a situation to be monotonous. Everything becomes routine, and routines kill creativity. That's why it's important to take periodic breaks.

Take a nap. Some of my friends make fun of me when I say, "I've got to take a nap." But you know what? When I wake up, I'm more charged, more creative, and more positive than I was when I woke up in the morning. Some people say it's cultural and some call it lazy. But I say it's the best way to continue the day and break a monotonous routine.

Another way to break a monotonous routine is to exercise. Get up and move.

We all have troubles. If you can't talk to people about your concerns, take up a sport. Go for a run. Healthy diversions are essential, and this is why we made riding motorcycles, especially those from Soriano Motori, easy, quiet, and healthy. You get to be yourself. You can hit the road, get away from everybody, avoid the virus, and be free. It's the perfect way to get away from problems and maintain a healthy mindset.

Another thing, limit the blue light. Throughout the pandemic, there's a good chance that people spend more time watching television, playing on their computers, or looking at their phones. That's not good for the eyes or the brain.

If you read scientific reports, you know this. The screen generates blue light and suppresses melatonin, which is a very important enzyme that boosts creativity. Studies show that blue light negatively affects sleep and circadian rhythms. That produces negative effects on physical, emotional, and cognitive health, which affects health overall. So try to limit your exposure to blue light.

Go outside, get some fresh air. Go to the beach, if you can. Head to the mountains, reconnect with the earth, spend time in nature. That's the best bounce-back strategy in the age of the pandemic.

The Exit

For the entrepreneur, the conversation about sustainability is about whether or not you want to keep your business going. If so, for how long? You have to think about your exit. What does it look like?

It's not clear what a sustainable future looks like for anyone. Nevertheless, with emerging technologies, and even improvements on older and cleaner fuel sources, many people today are looking to a post-fossil fuel world. That includes businesses. Even your own business.

Soriano Motori is building a global electric vehicle platform. Since the 1950s, after World War II, the US experienced unprecedented growth. We saw an increase in intensive farming, a technology revolution, a massive increase in the power grid, and we've put an incredible strain on the planet's resources. We're also far more aware of the plight of the developing world. There are natural and human costs to this economic growth and the effect that it can have on the earth's ecosystem and the human population.

We must develop new, cleaner technologies to cope with our demand because sustainability is not just about the environment. One of the biggest social activism movements related to social development has been the growing advocacy for sustainable and equitable trade relationships. This movement, championed by nonprofit organizations such as Fairtrade and the Rainforest Alliance, encourages good farming practices while ensuring farmers producing luxury goods, such as coffee and cacao, receive a decent living wage. Activists and sustainability professionals are hoping to remove trade barriers so that everyone contributing to world economies can benefit.

We all have to adapt to emerging technologies. Whether we do it to reduce costs or because it's the fastest way to transport people across the globe, we can't rest on our laurels. We've got to be aware. This is one of the greatest issues facing public companies, companies listed on the stock market. For that reason, a valid exit strategy is to build a brand appealing enough to bigger companies that they buy you out. You can take a break, go on vacation.

For private companies, there are typically two types of exits: go public or sell the company. Another option is to grow the company large enough that you can buy smaller technologies to make your company more competitive. There's built-in flexibility, and flexibility is important because you don't know how the future will impact your business. You can make educated guesses on managing growth based on statistics, market trends, and consumer behavior, but you can't predict black swan events and other external factors that might impact your business down the road. That's why you need to be flexible and not get locked into any one exit strategy.

If you're a tech company, an IPO would position you to be listed on a stock market, most likely the NASDAQ. If you're a manufacturing or transportation company, you could end up on the New York Stock Exchange. Once you go public, it's much more tedious work because there are rules for reporting your performance daily. It's like being watched by Big Brother. If you remain private, you have more control over how your information is relevant to partners and investors, which can be managed quarterly.

There's a natural dualism at play between public and private. There are other things you can do every day besides report to a government entity. You've got to think about that.

The IPO process requires money because it also implies legal and accounting firms. They are not cheap. You have to have big bucks in public.

If your exit strategy is to build a brand that is sexy enough, with strong leadership and a great bottom line, you can go to market and target big buyers just like you would with any other product. Your company is the product. If you find a buyer, you can sell it and take a break. Shut the engine off, relax. When you turn it back on again, you'll be more creative and you can come up with another great idea.

Another decision you'll have to make is whether you bring in partners. If you do, will they be equity partners, or will you use debt financing for the initial phase of your business? When you take on debt capital—meaning a loan, a loan equivalent, or a convertible note—you've got to pay that money back. If you can retain your equity, that's good because once you pay back your loan, then you're done. You can say goodbye to the financier.

Financing a business is like buying an asset, like a home. You take out a loan from a bank on a ten-, twenty-, or thirty-year note, and it isn't 100 percent yours until you pay off that loan. If you miss a payment, they can take it from you. Financing a business works the same way.

A loan is a debt burden. Some people like that because they like the pressure; they work better under that pressure. For others, it destroys their creativity because they have to come up with the bucks.

The other funding method is equity financing. You go to your investors and say, "I have this idea. You be my partner and I'll give you equity in my company. Anything we do together is long term."

Don't think short term. You're not the next Facebook. You have to be better than Facebook.

If the client, partner, or investor is the right person—a person with vision, a mission statement, and the same objective that you have—then you have a good person on your team. Your long-term exit should be an IPO or M&A transaction. If you understand that from day one, you'll have a fighting chance. But you can't make guarantees. You can't make promises. If I had a crystal ball for every venture I've had in my life, I'd be in a different place right now. But, obviously, that's not the case. Don't make promises you can't keep.

In other words, make sure you maintain realistic expectations.

Charity is another great exit strategy. If you have a lot of money, you can create a charity or foundation and give back to the community. If you have a large client, for instance, and they helped you grow, you can give back to that community in various ways. You can promote better education systems, build hospitals, support research for the benefit of people in the future, just to name a few.

My family has always been grateful for our business's growth, and I'll continue that tradition.

Some Terms You Should Know

Before you plan your exit, there are some terms you should know. These will help you better understand how certain internal factors can impact your exit.

Distressed assets: A distressed asset is anything you have to sell at a discount to acquire value. This might include debt, real estate, or any asset owned by the company.

Companies usually have distressed assets when they are going through financial difficulties. They may be facing bankruptcy or bad management that has caused their business to take a downturn and they need to unload assets to regain capital. This can be an opportunity for you to buy those assets at a discount to improve your company's own financial position.

Of course, if your company has distressed assets, it means you're in trouble. Having a lot of distressed assets sometimes leads company leaders to sell the company or make an early exit.

Let's say you do everything right and you sell your company for a lot of money; the first thing you have to do is pay your creditors. You have to pay off your loans if you have any. Second, you pay your partners. After that, you pay employees. Paying yourself comes last; that is prescribed by law.

During the financial crisis in 2008 and 2009, there were a lot of companies that went under, and a lot of people were upset with CEOs who were getting paid bonuses while firing a lot of employees. They put people out of work. You want to avoid that kind of situation.

If you've had people working for you for a long time, contributing to the growth of your company, then you want to take care of them. It comes down to being socially responsible and having good leadership skills. Many companies lose sight of this once they have financial difficulties. People have lost money, jobs and tainted their future due to bad decisions while in management.

When you execute a good exit, you need to take care of the people with whom you have worked. That includes investors, partners, and employees. They took a risk with you, so reward that.

Insolvency: An insolvent business is one that hasn't done well. It may be facing bankruptcy, the company's products and services weren't appealing to the market, or management didn't take the right approach to create a sales team that can reach out to new customers effectively. I've seen plenty of companies fall into insolvency.

Insolvent companies are ugly to banks. They'll never give you a loan because you don't have enough revenue and too much debt. It's financial mismanagement.

You don't ever want to become insolvent, but if you do, you want to take every legal means necessary to pull your company out of it. Restructure the business, if necessary.

Many factors can make a company insolvent. For instance, you could have the best hotel in the world and a pandemic comes along to disrupt your business, forces limits on travel, government shutdown mandates, and social distancing guidelines. In essence, you're forced to shut down the business because you don't have the liquid capital to ride it through. That doesn't mean you were unsuccessful. It doesn't mean you had poor management or had a lousy product or service. It simply means external forces suppressed your business.

In that kind of situation, you may need to restructure. Claim bankruptcy, restructure, and relaunch when the market picks up again. A lot of that is happening in the real estate sector and the hospitality industry.

When you restructure your business, you have to refinance your debt. Right now, the cost of bank loans is at its lowest. But you want to maintain good credit for as long as you can. If your business is insolvent, you'll have to personally guarantee your loans, and restructuring the business can help you out of the situation.

Federal bankruptcy laws are easy to research. Go on the Internet and you can learn about business restructuring. You can find the reasons for legal reorganization, how to execute a bankruptcy liquidation, and more. Maybe you're sitting on enough cash that you can go to bankruptcy liquidations and buy hotels at a discount. That's something Soriano Motori is doing right now because there are a lot of businesses becoming insolvent. They are desperate to get out of business because they cannot run it. It's happening to a lot of restaurants.

It's sad, but there are a lot of mom-and-pop shops in Europe going out of business.

You don't have to be a lawyer, but if you run a business, you should have a little lawyer inside of you because everything has legal implications. Bankruptcy laws came about in the late nineteenth century, but they've evolved over the years. It's good to understand how these laws affect businesses going through insolvency. If you have a good understanding of these things, when you do need to hire a lawyer, you can tell them what to do and not wait for them to tell you what to do. It's a more powerful position and can save you some money.

In order to prevent insolvency, you should save money for a rainy day. Most people don't do this, even corporations, but it's just as important for a business to save for a rainy day as it is for a family. Many start-ups hit the market, have a high debt burden, and a more established company in the market is selling more products. That's a common mistake. Grow your business organically, step by step. Being a disruptor doesn't mean you have to spend a ton of money. Demand will create itself.

Grow Your Business Organically

If you put to market something that is appealing, it will organically attract people to it. You don't have to do anything. It will sell itself.

I'm a big believer in deliverables. If you create a product that speaks for itself, you shouldn't have a problem with being financially responsible. Being a disruptor doesn't mean you have to spend all of your money to create value for an investor or partner.

Even during a pandemic, you can grow organically. There have been a lot of businesses going under during the COVID pandemic. Many of them could restructure, learn from their mistakes, and start again. Any crisis will give rise to new blood, new opportunities, and new ideas. After the pandemic, in the post-pandemic phase, you'll see some of these come to fruition. These are a part of the ten paradigms I discussed earlier, and some of them are already happening.

Remember the paradigms? Peak globalization, climate change, robotics, smart everything, space, moral capitalism, techcelerator, demographics, quantitative failure, which is government intervention, and recession.

Raising taxes on everybody won't work, but entrepreneurs can use that situation to launch an idea. If you can see the opportunities, your path to success is much more probable.

The global population is nearing eight billion. The world population is growing exponentially. Out of that eight billion, half of the people are already connected to the Internet. Many of them have multiple devices and all of them are connected to the Internet. By 2030, that number is going up from 4.5 billion to 7.5 billion.

Imagine that. To be a disruptor, you need an Internet-based sales strategy.

We know there are close to thirty billion devices connected to the Internet right now; that's more than three times the population. That number will go to five hundred billion devices by 2030. We have close to 2.2 million robots connected in manufacturing, which will rise to 20 million. Not only will our population grow in the next decade, but the number of robots taking people's jobs is expected to grow. More people will be competing for that one job in the office because one robot can do the work of several people.

More people doesn't mean more jobs. Technology reduces the number of jobs. It does mean more competition and to be more competitive, you need better education. Therefore, education is going to play a huge role in how people prepare for the future.

The number of electric vehicles will grow from two million today to thirty million in ten years, and more than 70 percent of American families will not have a primary residence in the US. They will move to Latin America to live and work in the US. They'll have a second home in the US, but it won't be their primary residence.

The reason for that should be obvious. If you make a lot of money working in Los Angeles but you live in Mexico, you'll be ten times richer in Mexico on the same amount of income. That is because of the cost of living; it's all about finding the economic standard that you can afford in a place where things are cheaper. More than one million Americans already live in Mexico; that's their primary residence.

The world is changing. Entrepreneurs need to be aware of these changes and understand the numbers. When you put together your business plan, your next idea, you should have a basic understanding of the statistics, the numbers that define the marketplace and the living and working conditions of the future. If you have a good idea, it will grow organically because of demand and the rise in the human population. Position it well, and you'll be successful.

Paradigm Seven: Moral Capitalism

We've entered the 2020s with capitalism focused purely on profit maximization on the cusp of reform—as it shifts away from shareholder supremacy toward greater involvement of stakeholders, that is, moral capitalism. The 2020s will mark the key decade to meet the UN Sustainable Development Goals (SDGs) set for 2030, where we see ESG/impact investing, quantification of values, and intangible assets disrupting the traditional investing approach.

United Nations Sustainable Development Goals (SDG) for 2030

Source: UN

Impact investing millennials are more likely than older generations to believe that better performance is possible through impact investing (source: US Trust Wealth and Worth Survey 2018). For asset and wealth managers, this suggests a growing need for ESG, thematic, socially responsible, and impact-first solutions.

We believe the trend toward thematic and ESG investing will intensify once Generation Z comes of age and is embedded in the productive economy. Our US Equity and Quant Strategy team, led by Savita Subramanian, estimates that US$20 trillion of AuM will go into ESG strategies over the next twenty years—the size of the S&P. The asset owner mix is changing—there is strong demand for investment decisions that consider thematic issues. This is especially true among Gen X and Gen Y (millennials), who are increasingly becoming the swing factors in terms of new asset formation.

Currently, 4.4 billion people in Generation Y and Z account for 59 percent of the world's population and US$21 trillion of income, and they are focusing more on thematic investing. By 2020, millennials will account for 16 percent of global private wealth. Eighty-seven percent of millennials believe ESG factors/

impact investing is important in their investment process (source: BofA US Trust Survey 2018).

The majority of millennials are driven by the belief that businesses should have ambitions beyond profit, with almost nine out of ten (87 percent) believing that "the success of a business should be measured in terms of more than just its financial performance" (source: Deloitte 2016).

BofAML Themes by UN SDGs

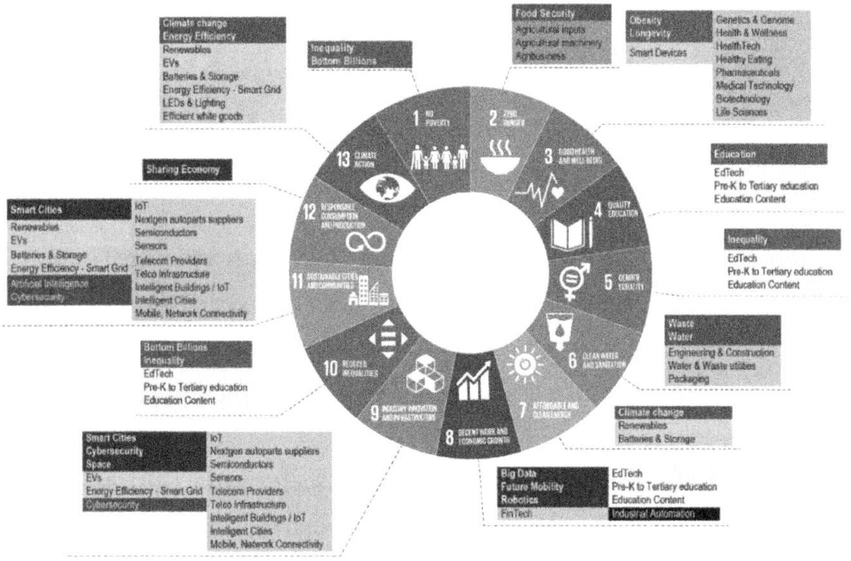

Source: BofA Merrill Lynch Research, UN

Sectors most impacted: financials, resources, REITs, transport, and utilities: TCFD's status report shows 50 percent of banks, consumer, and materials companies now disclose climate metrics and targets in ESG reports.

Highlights from an initial analysis of banks: 68 percent of exposure is indirectly through clients. UNEP initial scenarios assess the probability of default in utility increases 2.2–2.3x, and in metals and mining by 1.4–2.0x. Banks, including DBS, HSBC, and Westpac, now disclose exposure to "climate risk" sectors.

Insurers: risks including higher claims and reinsurance premiums. DNB estimates higher claims by 2085. Travelers provide probability ranges and losses for catastrophes.

Oil and mining: impact from lower demand (30 percent cut) oil prices and carbon price.

REITs: climate risk impacts REITs with coastal exposure, including across Japan, Hong Kong, and Singapore. Buildings will need to cut CO2 emissions 80 percent by 2050.

Transport: Aurizon says global coal trade could decline by 33 percent. China Airlines highlights that the abnormal climate leads to flight cancellations. Toyota notes cost and demand risks from fuel consumption regulation. Utilities: electricity share to grow from 20 percent to 50 percent but 90 percent cut in emissions.

Sheconomy: equality for women could boost global GDP by up to 31 percent. According to the World Economic Forum, the overall global gender gap could close in 108 years for the 106 countries the report has covered since its initiation. The most challenging gender gaps, economic and political empowerment, could even take 202 and 107 years, respectively. If current rates were to be maintained, the overall global gender gap would close in 61 years in Western Europe, 70 years in South Asia, 74 years in Latin America and the Caribbean, 124 years in Eastern Europe and Central Asia, 135 years in sub-Saharan Africa, 153 years in the Middle East and North Africa, 171 years in East Asia and the Pacific, and 165 years in North America.

CHAPTER 9

Delegation: How Many Hats Can You Wear at a Time?

You can't wear all the hats in your business. In fact, most entrepreneurs have to relinquish power once the business has matured enough to run itself.

That's easier said than done. One reason for that is that entrepreneurs start with their own idea. If they have the idea, the only person they trust with it is themselves. They develop the vision and mission statement and that becomes the way the entrepreneur would like other people to think about the idea. What happens, typically, is the entrepreneur grows the business to a point that it can run itself. Then they have to trust others.

Trust comes from building relationships. It comes from experience and communication.

As you build your team and communicate your idea, you have to make sure you allow your employees the freedom to reconvey information back to you so that you can align your thoughts and leadership style to the way you want the company to run.

Another reason entrepreneurs have that problem is they think their idea is the only one that will succeed. But disruptive ideas speak for themselves.

Ideas are internal. How do you take an idea and externalize it? You turn it into a tangent. You have to prove the concept. You have to send that idea on a journey and show people, after it has been completed, what you've achieved with it.

Ninety percent of start-ups fail, and it happens for this very reason. Entrepreneurs try to wear all the hats, but you can't. You have to find the right accountant, somebody who can help you set your books right. From day one, you have to have your accounting, your tax filings, and all of your finances in order. Don't incur unnecessary expenses. If you've been eating ramen noodles, keep eating them until you can move up to real food. You'll eventually get to the point where you can afford a real meal. You don't want to spend money unnecessarily.

Entrepreneurs get their businesses going and immediately start flying first class. You haven't flown first class in your life and all of a sudden, you're flying first class on somebody else's money. Don't do that. Wait until you have the revenue to justify first class.

Another flaw entrepreneurs have is pretension. They pretend they have a lifestyle they've never had before. When you do that, you're unnecessarily incurring expenses. Your burn rate is extremely high while not creating revenue. Investors will see right through you. If they trust you enough to invest their money into your business, they will be watching and they will see through you. So, avoid falling into that trap.

Also, avoid distrust. You have to build a team so that, if you get sick or something happens in your family that takes you away from the business, they will continue running the business for you. They represent you when you're not there.

Building culture from within the company is very important. You want your employees to represent you the same way you'd represent yourself.

It starts at the beginning of your business. You build an intimate relationship with your employees in that early stage. Find a lawyer that you can trust. A lawyer that bills you for every email you send, every second they talk to you, every text message, is not a good lawyer. You want a lawyer that will spend time with you and your family, explain the legal issues you can incur, and stop you from doing something you shouldn't do, something that might hurt your business. Also, when you're working with lawyers, make sure you know more about stuff that you need to do before they do.

Just like when you go to the doctor, you can learn all about diseases and treatments on the Internet. You can learn so much before the doctor tells you what you have or what you think you have before a diagnosis. It's the same way with running a company.

You've got to perform market research, stay on top of markets, watch trends, and track daily patterns of consumer behavior. The moment you stop doing that is the moment you're going to stop adapting your technology to advance your service toward disruption. So don't stop. Don't get sidetracked by a glamorous lifestyle. Don't think you have to pretend to be successful when you are barely off the ground. Bring your company to revenue as soon as possible.

When It's Time to Let Go

There's no set time. Every market segment behaves differently, but the first year is going to be the toughest because you'll want your business to scale. You'll want to go from zero to sixty in three seconds.

You'll want to communicate with your investors every quarter. Produce a quarterly report. And release a full report annually. You'll want to inform your investors on what assets your company has, what the company has accomplished, where you're going next, and what your milestones are for that time frame. Your reports show there's real, tangible, intrinsic value to your original idea. Your investors will see your growth. If you have their support, your achievements will be the first sign that you're doing it right.

If you have an investor who wants out, there could be several reasons why. If they want out simply because you didn't meet a certain milestone or benchmark, that's a sign you need to rethink your strategy. Start working with your team differently.

Delegation is critical. It should be automatic after a year that you delegate to your core management team. You need to concentrate on bringing more capital to the table. If you're the founder, let your team run operations and logistics. They have the experience and are better at it than you. What I do is hire people who are at least ten or twenty years older than me and ten years younger.

I learn from my team, even the younger ones. I make sure I'm not the smartest guy at the table. Some entrepreneurs think that when they walk in the room, they have to be the smartest one in the room; that's not true. Be humble. If you don't know something, then you don't know it.

You do have to be prepared. If you don't prepare for it, don't have the meeting because you're going to blow it up. Also, don't fall into bad habits, like staying out all night. If you stay out late, drinking, smoking, and doing crazy stuff, that will take your energy away. You must keep your mind rested. You should always be producing mentally.

There's plenty of scientific research that says when you overextend yourself or you don't get enough sleep, you won't function at 100 percent. You won't be effective at running a start-up with somebody else's money. Be disciplined.

Investors like to go out and have fun, that's fine, but you have to be disciplined. Tell them, "I'm going to bed because I need to make money with your money." They'll appreciate that more than you are partying or getting drunk and spending their money on things that won't earn a return.

What Should You Delegate?

Start at the beginning. Work more than you ever did. When you start thinking you need an assistant or a secretary, that you need people below you, you'll just waste money. You can take out your own trash, make your own food, and pick up your own loads. You don't need people below you to make you feel powerful. If you start out doing things yourself, by the time you hire somebody else, you'll be able to show them how it needs to be done.

Understand that everybody has different personalities. Everyone has a different perspective about work and how to get it done. What works for you might not work for someone else.

Also, be culturally aware of people's backgrounds. Be sensitive to their feelings. We live in a new world and have to behave professionally from day one, then carry that out throughout the life of the corporation. Your behavior will define the corporate culture. Other people will pattern their behavior on yours. You can have a hundred thousand successful companies and if one of them develops a hostile environment for employees, that one is a bad apple. That says that each corporation has a different culture based on the personalities and behaviors of the people at the top. A corporation is similar to people in that each one has its own personality.

From the top down, your core management team should consist of people who bring their ideas to the table every day. Do you have an open-door policy?

I'll have a meeting at the beginning of the week and at the end of the week and tell my core managers to bring their brains. I'll say, "I need a new idea." Innovation. Disruption. Change.

Don't be afraid of change. Your initial strategy will change. There are multiple ways to get from point A to point B; nothing is linear. You need to understand that and let people make mistakes. There will be bumps on the road. Learn from your mistakes and don't make the same ones twice.

When you hire people without experience, you're going to have to take the time to train them. You will lose time, but when you hire people with experience, they'll bring solutions, and you won't have to teach them how to do that. You'll spend less time training.

None of us are islands. That time you spend with someone who just graduated from college will teach you about new ways of consuming. At the end of the day, we're all selling a product or service. That new generation, the one that recently graduated from college, will be your future customers. So, you need to learn from your employees at the same time that you're trying to help them understand how the business should run. That's a win-win situation and you won't be wasting your time.

Don't look at things negatively; keep a positive attitude. I have interns every year from all over the US and Europe. I do that because I need to understand how the market is behaving. I want to understand their way of consumption. I examine the market trends and young people's buying patterns.

It's targeting the future of my company by learning from the interns directly. Teaching somebody to make coffee should be fun. Develop a light heart because students today do not get that kind of training. They learn directly from books and the success of others.

We talk about a sense of entitlement; that's a part of the educational system. As founders and entrepreneurs—it doesn't matter what sector you're in—we should be teachers. If you have people who are older than you, you're going to learn from them; that happens when you start with people who have already done what you're doing. You'll have people on your team with experience and with networks you need to tap into for distribution. That's why it's important to trust, and that's why you need to build relationships from the top down. Create win-win situations.

Before you hire employees, sign up for a payroll system. Many of them allow you to run a company. Investors need to see that development, and they want transparency. Fully integrated systems give you a better sense of building a real company. Consultants not focused on your product or service because they're thinking about consulting for other companies aren't going to help you. But if you want to build your own DNA, your own blueprint, then you need to adopt a payroll system, insurance, and medical coverage. They can be tricky because they're expensive, but you can manage a lot of it yourself these days.

Still, you have to be careful how you do it. Consultants will come to you with high technology, and they're not cheap. You need somebody on your core team handling financials and accounting for payments. I built my core management team during the first twelve or eighteen months by making certain people consultants. I have them sign an NDA; you've got to make sure you protect the information they have access to because they're tasting your secret sauce.

Your company is going to be an advantage for them. As your company grows, they will grow. They'll feel a sense of ownership as a consultant. If they find a better job, then they'll leave you. For that reason, you want to build your core management team as soon as you can. Investors will appreciate the fact that you are building a legal entity where the numbers are black and white. As you go, you'll learn better ways to reduce your costs, manage your burn rates, and so on.

Why Building Company Culture Is Important

A star in a start-up? That's kind of like a river with one water droplet claiming an elevated status. Like that water droplet is responsible for the river's success. It happens on its own in the very beginning of a company when you have just a few employees. Maybe you're the star.

But you have to get beyond that stage quickly.

People have set traits. Their personalities add to the team's personality, and the team personality contributes to the DNA of your company. You have to hire people who are professional and respectful of each other's ideas. You need people who are multiculturally aware.

Another thing is, people with experience come with a certain level of authority. If you're a founder and you're younger than they are, it may be challenging at first. Just tell them what you need them to do. To them, you're just a paycheck, but that's not what you want. You want them to see you as a leader. You want them to see you as the person they want to see every day when they wake up and hang out with when work is done. You want them to want to learn from you.

It goes back to being an educator. You need to challenge your team. Be that person people seek after for deep insights.

You can create a corporate culture through sports activities, competitions, idea sharing exercises, and using a think-tank approach. You can host charity events to show your people what you care about while learning what they care about and supporting them. Most people have some kind of interest, whether it's charity and foundation work, sports, movies, or whatever. Connect with that. This is what we call KYC (or Know Your Client). Be a part of their lives, learn about their families, learn about who they are and what's important to them because they're walking away from their families every day to be with you, to bring value to your company, and help you grow. So this is important.

If you're a founder or a leader, your employees should be in the back of your mind. Get to know their names. If you can manage to learn their birthdays or important holidays, send them a greeting card and they'll love you for that.

How you create loyalty in your company is important. Your employees, even if they get a better offer from another employer, a bit more money or better benefits, they'll stay with you because they like you and the company culture; I've proven that over and over. Before you even think about numbers, consider the social, psychological, and emotional aspects of business. Investors will see that you're an inspiring person and that will inspire investors too. You can change lives with the proper perspective.

If you think that building a culture is hard, you're right. It's like going to a game and expecting to lose; it's an attitude. You've got to have an aptitude for success, mentorship, and management. Don't be a defeatist.

You want to inspire your team to not be defeatist either. You can't do that if you're negative and thinking it can't be done. Go in expecting to create a win-win situation. If things happen that you didn't count on, learn from your mistakes and move on. Strategies will always change. Welcome that change and welcome the risk.

When we hear the word "risk," we tend to think it's negative. We start to look for a rescue team to come and help us. We think things are out of balance, but risk isn't negative. It's a necessity.

Always be ready. Every successful enterprise in the world today has a business continuity plan (BCP). Develop your BCP. You don't have to write it down but know what it is in your mind.

What we learned from this pandemic is that the businesses that survived are resilient; a lot of businesses got hurt. But many people continued living, working, thinking, and making things better. They adapted.

The entrepreneur should sit back, look around, study their surroundings, and observe how humans continue to move forward. The COVID pandemic is the first event like that most of us have experienced. When there's a need, when there is mass distress like we've experienced recently, people come together. 9/11 showed the world how quickly humans come together to help each other. The pandemic was the first event we've ever faced that keeps us apart. If we came together, we'd be contagious and pass sickness around to each other. But we found ways to connect people, nonetheless. We always find ways to keep going because we're resilient.

Corporations should have this resilience built into their DNA. If you make a mistake, a pandemic happens, a disaster occurs, you'll have your BCP in place. People will continue to function. The world keeps turning. Build a culture from within the company and it will survive hard times.

Going virtual is a big decision. What's tricky about technology that has existed for twenty years is that it evolves. We went from texting to holding virtual meetings. We had to reconnect on Zoom to continue talking about business. One way to instill culture is to also talk about fun stuff when you have that virtual business meeting. When we're in the office, we go out for drinks afterward, go out to dinner, and have some fun. So we should continue to do that in the virtual environment. Have a virtual cocktail party.

Everyone has a different perspective on the pandemic and how it touched their lives. That's why it's important to maintain the corporate culture. It allows people to lighten up, have fun, and tell their stories.

Maintain your sense of humor and throw a virtual cocktail party at your virtual office.

I think people got more personal in the virtual environment. The pandemic moved people to rely on technology, but it made their connections more personal. They continued to maintain a high level of connectivity. And now that we can get face-to-face again, people feel closer because they were able to connect online and in virtual environments.

That's what I mean when I say people tend to come together in times of distress. After the pandemic, as our markets, our society, and our communities open up again, people feel close because they didn't abandon their relationships. It's a change, but we must embrace change. Embracing change as it happens builds positive energy and that's a good way to restart after the disruption.

Delegation is about getting the business to run itself. That's true freedom for the entrepreneur because it allows you to create other ventures and chase other business opportunities. This is the activity that takes a lot of entrepreneurs out. Like I said earlier, 90 percent of businesses fail because their founders aren't able to delegate, and it may be more than that. If you can delegate, you can be successful and do it over and over again.

I like to read. While reading a book on the entrepreneurial mentality, I read about a survey in Europe. They went to a university and asked this question: What kind of job would you like to have when you graduate? More than 90 percent of those surveyed said they wanted to work for the government. The same survey was taken in the United States and fewer than 10 percent surveyed would even consider working for the government. They wanted to be their own boss.

When you want to be your own boss, when you want to implement your own ideas, that demonstrates a big difference between cultures.

Independence breeds freedom in the markets and allows disruption. Therefore, delegation comes with wanting to be yourself, to be an innovator, a pioneer, a leader, and a person of influence. I'm not talking about social media influence. I mean influencing people by the way you wear your shoes or your belt. You can make people feel good about themselves and inspire them toward success.

Inspiring others is a strong drive. It drives loyalty in the mind, loyalty to the leader, the product, your presence, and membership in your organization.

Delegation is about inspiration. Do it well and you'll build loyalty to your company.

Paradigm Eight: Splinternet

The race for technological supremacy and sovereignty Internet that we've entered in the 2020s with the technological rivalry between the US and China heating up, that is, splinternet. There could potentially be a bifurcation in tech standards by the end of the decade driven by a divergence in the rules and standards of the superpowers ("sovereign internets").

In the next decade, we will continue to see the US and China competing in NextGen technologies like quantum computing, semiconductors, AI, 5G communication networks, cybersecurity, and space that go well beyond manufacturing and trade issues.

By 2030, China is expected to overtake the US and become the world leader in AI. Furthermore, China's One Belt One Road project already spans sixty-plus countries with a total of 4.4 billion people, covering 63 percent of the world's population.

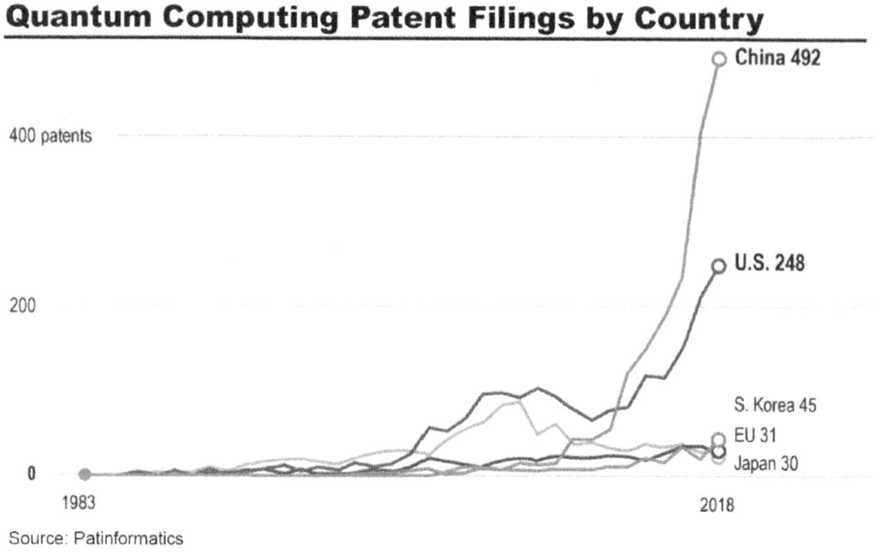

US Versus China: From Trade War to Tech War?

We believe the current trade war will transition toward a tech war in the 2020s, which will see a new "arms race" between the US and China to reach national superiority in technology over the long-term vis-à-vis quantum computing, big data, 5G, artificial intelligence, electric vehicles, robotics, and cybersecurity, and so on. China's strategy is to ensure that 40 percent of its mobile phone chips, 70 percent of its industrial robots, and 80 percent of its renewable energy equipment are "Made in China" by 2025. This China First strategy will be met head-on by an America First strategy.

The IMF forecasts military spending by the US and China to rise substantially in the coming decades to US$3 trillion and US$4 trillion, respectively.

Global defense, tech, and cybersecurity as "national security champions" should benefit over the next ten years.

Moreover, the US tech sector could be inhibited by increasing "Occupy Silicon Valley" sentiment and regulation policies to address inequality. Revenue-rich tech is the most lightly regulated sector: just 27K US federal regulations, compared with 215K for manufacturing and 128K for the financial sector. And tech and e-commerce companies currently account for almost a quarter of all US corporate profits, a level rarely exceeded and often associated with bubble peaks.

Diverging data trends mean that BAT (Baidu, Alibaba, and Tencent) may be better placed than FAANG (Facebook, Amazon, Apple, Netflix, and Google) to take advantage of AI and big data developments as key megatrends drive further potential growth in China.

Just over half the population is connected in China but that is already nearly triple the number of Internet users in the US, suggesting China's annual mobile data traffic could grow 56 percent compared with 35 percent in the US (source: Internet World Stats, Cisco, World Bank). With favorable policies and government backing, China's technology companies are likely to be better placed to take advantage of these data trends.

Data Privacy: From Big Data to Big Brother

	China	US	Europe
Annual Mobile Data (Exabytes 2016)	11.8	16.1	8.8*
Annual Mobile Data CAGR to 2021	56%	35%	42%*
Connected people	772mn	292mn	705mn
Unconnected %	45%	11%	15%
Data policy	Data control	Data commercialisation	Data protection
Legislation	Cyber Security Law of the People's Republic of China	No federal law, California Consumer Privacy Act (2020), Illinois' Biometric Information Privacy Act	General Data Protection Regulation (GDPR)
Privacy concerns	Limited	Heightened	Heightened

Source: Internet World Stats, Cisco, Control Risks, World Bank *Western Europe only

China has over 2.6x the Internet users of the US and only 56 percent penetration. With 772 million Internet users, China has already overtaken the US and Europe with 287 million and 705 million, respectively, and has a significantly lower penetration rate of 56 percent (the US and Europe have 89 percent and 85 percent, respectively) (source: South China Morning Post, Internet World Stats). Consequently, the country has the opportunity to grow data by bringing half its population online.

China's data volumes are also expanding above average, with 21.21 billion gigabytes of data traffic generated in just eleven months in 2017 and mobile Internet access traffic increasing 158.2 percent (source: CNNIC 2018).

In comparison, Cisco estimates a 46 percent CAGR in global mobile data traffic to 2022 (source: Cisco 2018). China's first formal cybersecurity law came into effect on June 1, 2017, banning the collection and sale of users' personal information without permission.

China's Larger Pie

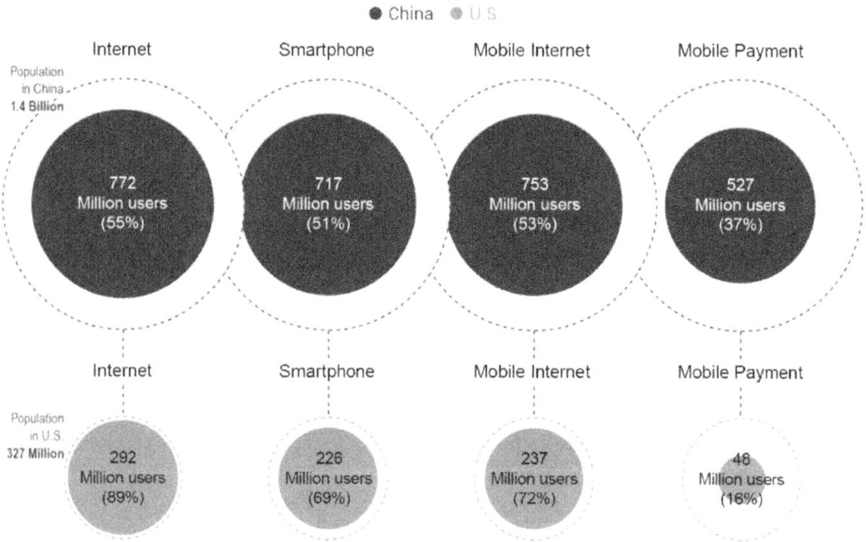

Source: Abacus News 2018

Companies will also need to store data on servers inside China with restrictions on the transfers of certain data of Chinese citizens out of the country by "key information infrastructure operators," including banks, power companies, and others.

Major multinationals believe this will give Chinese companies an unfair advantage as it prevents them from pooling client data in cloud storage centers across the world. Seemingly less privacy-conscious in China—surveys in the US and Europe have reported greater privacy concerns—of 2,000 surveyed in the US in November 2017, 82 percent were concerned to some degree (source: GDMA 2018).

China seems less affected by this issue. When questioned about the social credit system to collect data on all Chinese citizens, in an online survey of 2,200 citizens, 49 percent claimed to strongly approve of the system in 2018 (source: Genia Kostka, Freie Universität Berlin 2018).

CHAPTER 10

From Entrepreneur to Investor

Once you start a certain number of businesses and create your own vision, you start to see the value of supporting other entrepreneurs while leveraging your skills to invest in their businesses. At that point, it's a smart strategy to become an investor.

In order to understand the benefits of being an investor, you need to understand the differences between being an investor and being an entrepreneur. As a young entrepreneur, my mentors would tell me that I should always understand the *why*. Why should you understand the difference?

First, an entrepreneur focuses on a business operation. The investor focuses on the commercial and financial side of the business. As an investor, you have to be ambassadorial. Bet on the people, not the project.

Second, an entrepreneur comes up with new business ideas while the investor considers the existing business idea. New ideas tend to create confusion and the investor should say, "Nah, it's too complicated. Stick to one thing." That's the kind of relationship you want.

Third, an entrepreneur approaches an investor for a business financing opportunity. They'll say, "Here's my proposal, and here's my business plan," in order to show that they know what they're doing. On the other hand, investors will approach the entrepreneur if the business looks profitable. They're concerned about the bottom line.

Another difference is when the entrepreneur is passionate about the business and dedicated to their idea. Most of them are persistent and tend to stick to running the business even though it may suffer losses, but an investor is more practical to the commercial evaluation. They may leave the business if it loses money.

Entrepreneurs also tend to be optimistic about their presentations. You'll never hear one say, "this isn't going to work." They'll present the best-case scenario with hockey stick growth. The investor is not optimistic, typically, but they may be more optimistic if the business value is relatively high. It goes back to the numbers.

Another way investors and entrepreneurs are different is in whether or not they invest in the company. The entrepreneur doesn't always invest money in the business, but they'll be in the trenches running the operation. The investor only invests money. They're never going to work for the company; they don't get involved in the daily operations. This is why the entrepreneur needs to maintain communications with shareholders at all times. That might mean hanging out with them at the bar, ordering a pizza, or whatever. You need to have a connection beyond the numbers.

The entrepreneur reports to the investor regularly. It should be at least quarterly or monthly. I do it daily.

In return, the investor reads and evaluates the entrepreneur's reports. It may be advantageous to both that the investor makes suggestions or presents some ideas to the entrepreneur to improve the business. They may make connections for the entrepreneur, giving access to people and resources in their own network.

An entrepreneur can start a business with no money, but money is the main tool for the investor. The entrepreneur starts with an idea and has to find clients. You can do that without putting money on the table.

The entrepreneur will always know the business more than anyone else. They manage the daily operations, marketing, sales, dealing with employees and customers. But the investor knows only what they read in financial reports or whatever is reported to them by the entrepreneur. Their view of the company isn't based on dealing with people—customers, employees, and partners.

The final difference is risk. For the entrepreneur, the risks are energy and time. Sweat equity. The investor risks money and time.

An entrepreneur learns about business from scratch. After proving your concept and business model, and after making some money, you can then become an investor and offer strategic value. You become what the market calls "smart money." Entrepreneurs will always benefit from working with smart money because that investor can teach them, coach them, train them, guide them, and help them get their idea to market.

There are also benefits to transitioning from entrepreneur to investor. First is strategic knowledge. Secondly, you've built up your networks through years of business and can now tap into those for your next wave of investment, and then there are potential integrations. You may not want to invest in a company. Instead, you might want to add it to an existing business plan so that it grows into something with a global impact.

Get Some Skin In the Game

If you're a successful entrepreneur, you know what it takes to make sure your business is prosperous. You leverage the expertise of other people. Investing money is a business risk and that's why there is an entire industry of people who don't want to be experts themselves. For instance, with some of our companies, we have doctors, dentists, and other successful professionals who are not experts in engineering, design, or fashion. They risk their money by lending to or investing in other businesses.

That's more personal. It involves relationship development, so it's not just about the investment. They need to believe in what you're doing. The entrepreneur needs to prove that what they're saying is possible. That means they must be optimistic, persistent, and work hard to ensure the business is successful.

Investors don't have to be entrepreneurs. They want to invest money to get a higher return. An investor who invests to lose money is what we call "dumb money." You don't want that in your business. You want the smart money, someone who invests in you because they believe in you. The upside is, if you're successful, you'll make money together, but that's not your only objective.

As an entrepreneur, you can train yourself to be a better investor. You learn from your investors and that teaches you to be a better investor. After transitioning from entrepreneur to investor, you'll have seen both sides of the coin. Then you can be the smart money.

Becoming an investor means investing in yourself first. Personally, I've always invested in my own businesses. I provide the seed money. I make sure my money is the first money in. I put in the time and put in my brain, and, of course, my cash.

I do that on purpose because people will ask, "Do you have skin in the game?" They want to know how much money you put into the business. They're not asking how many hours you put into working on it. But the entrepreneur needs to translate their investment into dollars and cents. Did you go to your bank, draw money out, and put it into your business? Or did you count your hours, translate that into what you might have earned if you were working another job? The investor wants to know what your investment is.

Some entrepreneurs don't have money. An eighteen-year-old high school dropout isn't going to go to the bank and withdraw $100,000 at the drop of a hat. A twenty-one-year-old college student isn't likely to do that either.

In my case, I had worked in banking for twenty years. I had put money aside so that I could invest in my own ideas. That shows investors that I have skin in the game.

Today, I'm so comfortable about my investments that all of them carry my surname. The chances of failing at something with your name on it are pretty slim.

Investing in myself has always been a core value of mine. I don't tell everybody that they should do it, but if they can, then it's a good thing. If they can't, that's why there is a seed round. You go to your family and friends to help you get started. You have a much closer relationship with these people. Because they are your closest relationships, you will try to do what is best for them.

When you start going to strangers to raise capital, which we all do, you need to learn how to build those relationships. That provides a level of trust necessary for someone to give you money.

Every investor has a different approach to assessing what kind of investment your business might be. What have you done? Who are you? Who is your family? Where have you worked? Where have you studied? This is what we call due diligence. They also call it figuring your shit out. Investors have to know who you are.

Social media allows us to investigate people more easily. We can understand people's lifestyles. There is a lot of data out there that can provide us with a pretty good picture of the type of entrepreneur or the type of investment that we're dealing with. But you have to be careful with what you put out there. If you show on Facebook, Instagram, and TikTok that you're always drinking and partying, or doing crazy stuff, that kind of behavior isn't going to attract investors who say, "This guy is trustworthy" or "This girl is the best."

Some More Terms You Should Know

As an entrepreneur and an investor, you should be familiar with these terms.

Future wealth manager: The money that comes from other investors doesn't mean it's money that you've made. That's not your wealth, it's not profits in your books; it's working capital. If you put that capital in the market, through your business venture, it will generate sales. You create revenue from that. Out of that revenue, you subtract costs and that's your profit or loss. A part of your costs will likely be your own income. Over time, the company will grow. It will either go public or go through a merger and acquisition event, which means you sell the company and create more wealth.

As you create income, you create wealth. Money goes from your company, to you, to your bank. Now you're handling personal costs. From your income, you pay your cost of living and that's going to be whatever savings you have left in the bank. That's your wealth. How do you manage that?

Just like in your business, you have to define yourself. What are your objectives? Do you want to buy a home? Do you want to get married? Do you want to have children? And if you want to do all of those things, where? How? Why? You need to have some sort of family plan that allows you to meet those objectives.

In essence, you are your own wealth manager in the first stages of your life. If you become successful, you go public or sell your company, and you have $100 million in the bank, at that point, you aren't qualified to handle that money because you went from a salary of $100,000 to $100 million in your bank account. If you don't know how to handle that kind of money, you need professional help. That's why the banking system offers wealth management platforms.

On these platforms, a specialist comes to you and says, "These are the investments you should have based on your personal objectives." You can make investments into other ventures out of that money. But it always comes down to one question: How much taxes are you paying out of that? If you can't answer that question and you want to build a $1 billion company, you're going to have a serious disadvantage when you cash out.

It's a good idea to be your own wealth manager initially, but 99 percent of entrepreneurs don't do that. They can't. They don't have the time, and they don't have the passion for it. Filing IRS paperwork is boring. But you need to have somebody who can help you do it if you can't or won't. That will allow you to align your personal objectives with your business.

Again, there's a huge distinction between what your company is doing versus what you're doing in your personal life. Don't mix those up. You are not your company. Your company is a completely different entity. Treat it differently than yourself because that's the best way to become the future wealth manager of your own patrimony.

Alternative investments: If you took a finance or economics course in school, then you understand that the capital markets are two pillars of the US economy. The first pillar is debt financing, bonds. That means loans. And the second is equity financing. Equity financing is the most convenient tool for most entrepreneurs.

Debt financing happens in two different ways, either publicly through government entities or private lending. Private lending can include bank loans, lines of credit, home equity loans, school loans, and more. It's important to have a clear understanding of how money is provided to the entrepreneur because, one way or another, you're going to rely on the banking system.

Like debt financing, equity financing can be public or private. Publicly traded companies are listed on the biggest exchanges, such as NASDAQ, the New York Stock Exchange, Dow Jones, and so on. On the private side, what we call "alternative investments," equity financing involves new corporations, ideas, and innovations that don't involve the stock or bond markets.

By definition, alternative investments have the highest risk but also have the highest potential reward. When you compare stocks and bonds to alternative investments, there's a huge difference. On average, stocks and bonds can provide 8 percent to 10 percent returns. Alternative investments average 500 percent. That's a big difference.

If you invested $100 million and saw an 8 percent return, that means you made $8 million. But if you'd invested $100 million into an alternative investment and multiplied that by five times, you went from $100 million to $500 million.

Like I said, while alternative investments have higher potential rewards, they also have the highest risk. You could lose your $100 million as fast as you make $500 million.

Alternative investments are one of the paradigms I've defined for the coming decade. People are investing billions of dollars into them and seeing huge returns. Every portfolio should include some alternative investments.

For example, is COVID the catalyst that shifts autonomous vehicles (AVs) into gear?

One unlikely beneficiary of the pandemic could be the commercialization of self-driving vehicles. Until now, their commercial deployment has faced several challenges:

1. Regulatory approval varies internationally with no single standard or certification level as of yet.
2. Cost competitiveness and the technology deployed in the vehicles can vary, resulting in a current estimated cost of $150,000–$200,000 for car-based autonomous vehicles, according to our autos team, due to requiring specialized hardware such as LIDAR and new as-yet unproven software.
3. And lastly, the public acceptance of traveling in a vehicle without a human driver seemed a scary concept to most: just 22 percent claimed they'd feel safe in an autonomous vehicle in the Global Mobility Survey, for example.

The crisis may change some of this rhetoric and mitigate some of these challenges, accelerating the piloting of and eventually the transition to automated vehicles. We see altering development stages for personal and commercial vehicles.

Robotaxis: lack of a driver switches from negative to positive perception. The need to socially distance and limit human interaction could lead to a short-term preference for private over shared mobility and public transport. As such, driving a private car might be preferable, but still wouldn't be feasible in many big cities owing to congestion, parking, and air quality restrictions. Use of a commercial autonomous vehicle service for private use—a so-called on-demand robotaxi such as those being developed by companies such as Waymo, Cruise, Yandex and others—could be relatively more attractive and allay some of the fears that the public has around AVs. In essence—the main drawback or concern that the public had with AVs—the lack of a driver—could become its biggest selling point in a post-COVID world.

Regulation, technology testing, and scalability mean it might take longer. Despite the potential customer acceptance, the majority of companies testing AVs on public roads have had to pause operations to ensure the safety of their riders and engineers during the pandemic. In the US, for example, Waymo, Argo, Cruise, Zoox, Baidu, Pony.AI, and Aptiv all temporarily closed their robotaxi testing. This limits the data collection, mapping of streets, and eventual commercial launches of such services.

Regulatory approval also remains a key hurdle to the deployment of AVs at scale, as well as the need for certifiable technology. This has proved costly so far, with over $29 billion of disclosed investments into AV companies between 2010 and 2019 (McKinsey), as well as likely several more billions invested not disclosed by public automotive and technology companies.

Alphabet's Waymo raised a further $2.2 billion in March 2020, for example, indicating the significant investments required to develop the technology. The rising investment requirements coupled with potential limits to private funding available in the likely post-COVID recession may lead to some consolidation. Thus, AV deployment for passengers in the form of robotaxis is not going to be achieved quickly at scale, and in fact, may be slightly delayed in light of COVID-19.

Autonomous (commercial) vehicles: robots to the rescue. While most transit-based public autonomous vehicle trials were closed, those used for civil and delivery purposes have somewhat accelerated. Some examples of their deployment and changes during COVID-19 included robot vans deployed to disinfect streets.

For example, start-up Neolix deployed 200 vehicles in February/March 2020, compared with 125 units since their production started last May, including 50 vehicles across ten cities in China during lockdown periods for disinfection of roads.

- Autonomous delivery vans such as Nuro (delivering Domino's Pizzas in the US) and small delivery robots completed food and parcel delivery. Starship Technologies extended its pilot in the UK for food delivery, for example, and Yandex launched its Rover vehicle (Exhibit 10) in Moscow for city staff to receive documents and packages while offices are closed.
- Autonomous shuttles (e.g., Navya, Optimus Ride) were repurposed from transporting passengers to medical supplies and meals/packages, respectively.
- Long-haul autonomous freight pilots were expanded; TuSimple's fleet of forty autonomous trucks was deemed an essential operation in the US, delivering goods and food between Texas and Phoenix, and commenced free deliveries for a local food bank.
- The pandemic has highlighted the fragility of supply chains. Autonomous commercial vehicles could be part of the solution longer term, and their deployment could be accelerated based on the lessons learned during the crisis.

Exit strategy: It starts at the beginning of your idea. You raise capital, meet with investors, meet with customers, and make money. The investors want to know when they're going to see a profit after putting money into your bank account.

If I give you $1 million right now, what are you going to do with that? How's it going to make me money? You need to have an exit plan in place because they're going to want to get out when they reach a certain profit margin. As a company, we'll make money when it goes public or when it becomes sexy enough for a larger company to buy it.

There will be a process of evaluating the company, called an enterprise evaluation. The end result is an expected valuation after three, five, and ten years. That is what the investor needs to see. How are you going to get from where you are today to that valuation in three, five, or ten years?

Of course, anything can happen between the beginning and the exit. There's no guarantee; we've learned that in the last two decades. Starting with 9/11 and moving on to the Great Recession, then the financial crisis of 2008–2009, and now a global pandemic. It's hard to predict where things are going. That's why entrepreneurs and investors should consider catastrophes or challenges that might get in the way of business. Consider them, but develop a conservative mindset. You want the exit strategy to be realistic, and don't oversell it.

Instead of telling the investor, "We're going to have an exit of $1 billion," give them a lower number. If you're between $100 million and $1 billion, the investor will like the low number. If you exit at $1 billion, you'll be the golden child. But if you set the exit at $1 billion and only exit at $100 million, then you'll just be the copper child. So be careful with your exit expectations.

Systemic liquidation: The word *systemic* comes from finance. It relates to what we consider risk. Systemic risk and systemic liquidation. Remember this acronym: MIRI.

Market risk. Interest risk. Reinvestment risk. Inflation risk.

Those are the four types of systemic risk. A systemic liquidation takes into account all four types of risk and allows the entrepreneur to sell the company at the best possible value. Systemic means it will happen no matter what and those risks will always be present.

A market risk is when the market is not ready for the product or service, or the consumer doesn't have enough money to go into the market and buy the product or service.

Interest risk relates to the types of interest we have in our banking system. It starts with the federal rate, which is interest on loans from the federal government through the banking system.

In evaluating interest risk, take into account loan broker's risks, which is when banks loan money to other banks. There's also the discount rate and the prime rate.

These four types of interest rates fluctuate within the market. When we go to raise capital, we either go to a bank or to a private lender to get a loan. The interest rate represents your interest risk.

Reinvestment risk relates to whether an investment will earn more or less when moved to a new investment class. Let's say you have two investment opportunities. You could invest in a US Treasury bond today that pays 2 percent annually and matures in ten years, or invest over the same ten years in a convertible note that would pay 3 percent interest. Do you keep your money in the Treasury bond or move it into the private instrument?

Which one do you trust more? You know the government will always pay. But Joe, the alternative investment guy — you don't know what he's doing with his business. It really boils down to trust. Who do you have a better relationship with?

Reinvestment risk takes place at the time of decision. You have to decide where to put your money, but you also have to evaluate what will happen to your money in the next five or ten years. If your money is locked away for ten years and inaccessible, then you can't access it. But, after five years, that alternative investment you turned down could be paying 10 percent interest. There's your reinvestment risk because you made the wrong decision.

Inflation risk goes along with whatever the government is doing with money. When they print money or incentivize consumers to spend money. Any time there's a downturn in the economy, the government tries to incentivize spending by printing more dollars. Consumers feel safe and get into the market and buy things. That translates into the gross domestic product (GDP). Entrepreneurs should keep these things in mind when evaluating inflation risk.

Selling the Venture to Potential Buyers

Here's the dilemma a lot of entrepreneurs run into: pride.

You have your company; it's your baby. This is your first time. You put all of your sweat into your business and now have the opportunity to sell it. But who do you sell it to?

The second question is, can you be selective with the buyer?

Thirdly, do you really need to sell?

These are the three questions you're always going to face when selling a company, and it all depends on market parameters. It depends on how the market is doing. If the market is extremely bad and your company's doing extremely well, then you have a premium. You can see a beautiful exit. But, at the same time, you can potentially grow it even more and sell it later for more money. You have to make the decision.

That decision is one every entrepreneur will have to make. That involves setting up your personal objectives. Again, that is different from your company's objective, and investors will have their own objectives. You may not keep that in the back of your mind when running the company, but it's important to consider because investors want to exit with a profit. The exit is what they are expecting.

Understand your expected value. In any market, that's the classic mistake. There are buyers and sellers, losers, and winners. So, you have to be clear on your objectives.

If you want to be a selective seller, you should expect fewer offers. If you're extremely picky about who the buyer is, that will create fewer opportunities. That happens a lot. It happens most often when homeowners have a property they've grown up in, had children and grandchildren over for holidays, and created a lot of memories. Then they get to the point where they can't pay for the property anymore. They can't pay the taxes or maintain it properly, and no one in the family wants to live there anymore. Now you have a dilemma.

Do you sell the property to whoever? At sixty-five or seventy-five years old, do you just take the money and retire, or do you become the selective seller who tries to find the perfect family with the right background, the right last name, the right face, and the right occupation?

Most people don't care when they have to sell. If they have to sell, they have to sell. They just know they have to get rid of the property. It's the same way with companies. You can have the best company in the world right now, and the best product, but you're going to get to the point where you're going to either be the very best or the only company doing what you do.

Another question you'll face is, do consumers still use this product? If there's no competition and you are suddenly the only company in the market, chances are, you're going to die in the market because it's a dying market. That's something else to be aware of.

Selling a company is not as easy as it sounds. There are a lot of elements to calculate and account for. Be ready for that and consider all the market parameters around you.

Leveraged Buyouts and Managed Buyouts

Leveraging a buyout is more about merging or acquiring other companies. You can leverage your position by getting more investors involved once your company reaches a certain value.

Let's say you have a $100 million company. You can raise ten times that amount of money in the public market, or with loans. If you have the leverage, you can use the leverage. You can go to market, raise the capital inexpensively in the form of a loan, and use that capital to buy out other companies that will help you grow into a conglomerate.

In terms of performing, those sorts of leveraged buyouts (LBOs) are by far among the best exit strategies you can choose. It often happens with big tech companies. It's not that Facebook is creating or recreating innovation. No, it grew. It started leveraging and buying out smaller tech companies then integrating them into existing products. Now we have the monster that it is today.

There are a lot of them, too. Four percent of those monster big tech companies are US-based. They represent the fastest wealth accumulation that we have seen in a millennium.

If a company wants to go public, there are several factors to consider. There are advantages and disadvantages.

An advantage is access to capital. It's the fastest way to get investors because it's not just local at that point. You're doing the right thing, running the right company at the right time. And you'll raise enormous amounts of money that will allow you to grow your company in the direction that you want. Then you have to communicate with all of the public investors. You have to comply with rules and regulations, and regularly file paperwork.

When you're a private company, you can handle the business with less time focused on stress-related conditions. But it's a decision to make. Do you go the public corporation route or stay private?

Being in the eye of the public means people are going to talk about you. They may praise you or criticize you. They may say great things about you or talk poorly about you. Can you handle that? Remember when you were a start-up and you were super optimistic, when somebody said something negative about your business, did you get emotional about it? Did you respond with something like, "Who the hell are you? How can you say that?" If you do, you'll probably get into a fight with that person.

If your pride is hurt easily, or any sort of criticism of your idea, your baby, is painful to the extent that you lash out publicly, then you may not be ready to go public. You need to have thick skin and guts to go public and know how to deal with criticism.

You can't control what other people say about you. It can get ugly out there. Either you do one hell of a job and everybody's going to like you, or you're going to have all the greenies in the world telling you that you're not doing the right thing, or the mothers telling you that you're not supporting their community. Maybe the gay community isn't attracted to your brand. It can become political really quickly. You need to understand that before you make the decision to go public.

The nice part is, you get to be in front of the bell, you rang it. People will take pictures of you and put them in the newspaper. While you're in the public eye and getting all of that attention, you are also at the mercy of the public. You'll need a good PR firm, positive marketing, and a great advertising campaign to make sure you know how to fight the negative talk. That can be expensive.

Always keep the advantages and disadvantages in mind. It's easy to raise capital going public, but you're a target.

What An Investment Bank Does for Your Venture

Investment bankers are the writers of your business. They're a third party that validates and certifies, with their name and signature, that your assessment of your finances, your business operations, and your marketing approach is solid. You want them to help you put value to your stock before you go public.

If you authorize X amount of shares to raise Y amount of money, and those shares represent a percentage of equity in your company, that's what an investment banker will do for you. They're going to tell you how much of that equity, that percentage of your company, is worth in the market. Based on that, they can raise more capital for you publicly or privately. You need those instruments to be successful at raising capital, and investment bankers will be the ones to help you get there.

Paradigm Nine: Smart Everything

We've entered the 2020s with thirty billion devices connected to the Internet, but this figure could grow to five hundred billion by 2030 driven by the growth of smart cities, which includes future mobility, energy, 5G, IoT, among others. Urban areas already generate 85 percent of global economic activity but this is set to rise further from US$62 trillion in 2015 to an estimated US$115 trillion by 2030 (source: OECD World Bank). Rising urbanization in the 2020s (>1 billion urban population added by 2030 vs. 2015) means that problems inside cities will be magnified. However, new solutions such as smart mobility, smart buildings, big data solutions, infrastructure, and communications will be key in improving urban life.

Smart Cities

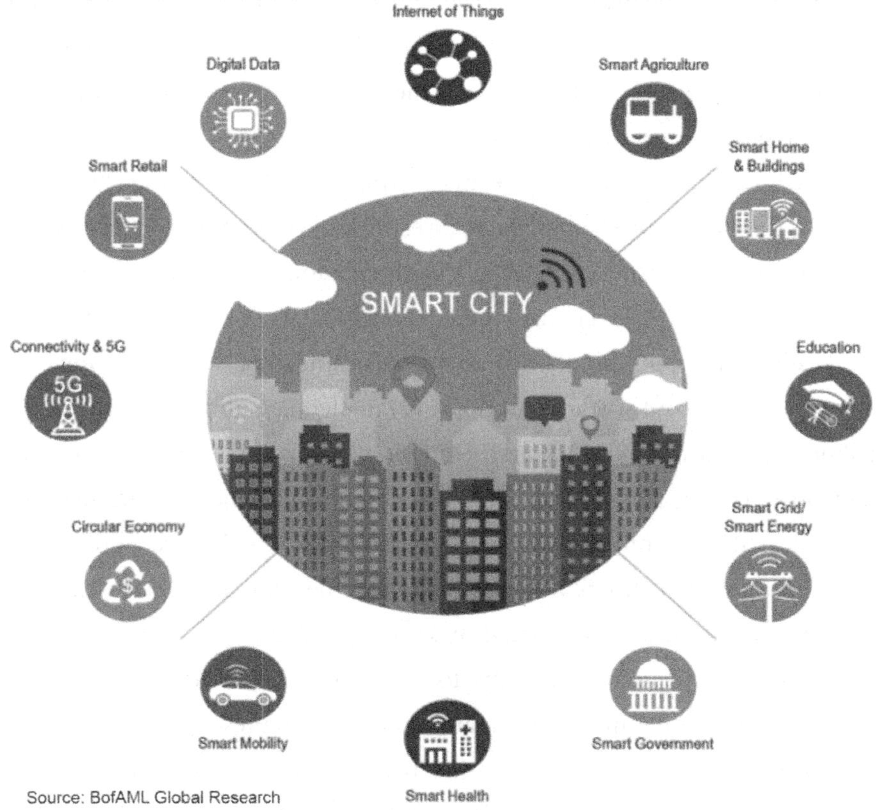

Source: BofAML Global Research

The "datafication" of everything. By the end of the decade, we will have five hundred billion connectable devices versus "only" thirty billion today (source: Cisco). By 2030, autonomous vehicles (AVs) could generate more data than the earth's entire population does today. One smart city will generate even more data than all the AVs combined. Three billion more people will go online, and we will interact with an online device every eighteen seconds by the middle of the decade—compared to six and a half minutes today. New technologies are needed to meet demographic, economic, environmental, social, and infrastructure challenges.

These include ubiquitous broadband (84 percent globally); NextGen infrastructure (5G up to 100x faster vs. 4G); the Internet of Things (IoT) (500 billion connected devices by 2030E); big data (175 zettabytes of data worldwide by 2025 per International Data Corp./IDC); cloud (secure, accessible), autonomous vehicles (fewer accidents) and artificial intelligence (predictive insights, anticipatory actions). Smart cities can achieve 30 percent lower energy use and crime, 20 percent fewer traffic delays and water loss.

Cities will drive some of the major transformations in the 2020s and beyond. The shift from a 55 percent to an estimated 70 percent urbanized world through 2050 will result in growing wealth creation, already at 84 percent of global GDP. However, less than 80 percent of global cities are showing fragility, or already at a tipping point with several issues needing smart solutions: (1) poor governance and weak institutions; (2) inadequate infrastructure (US$78 trillion required in next 10 years); (3) rising inequality (75 percent of cities are worse off than 20 years ago); (4) housing (1 billion new homes needed); (5) crime; (6) environmental challenges (cities use about 75 percent of natural resources and air quality is worsening); and (7) new and pervasive risks (e.g., cybersecurity, terrorism, security, disease, and pandemics).

Among the biggest ecosystems for investors, the smart city theme, by virtue of its broad definition and many subthemes, represents one of the biggest investible universes. Zion Market Research sees the global smart city market at around US$1 trillion in 2017 and expects it to reach US$2.7 trillion by 2024, growing at around 16 percent CAGR in 2018–2024. Mordon Intelligence values the smart cities market at US$0.5 trillion in 2018 and expects it to grow to US$1.45 trillion by 2024 (18 percent 2019–24 CAGR).

Smart City Digital Ecosystem Components

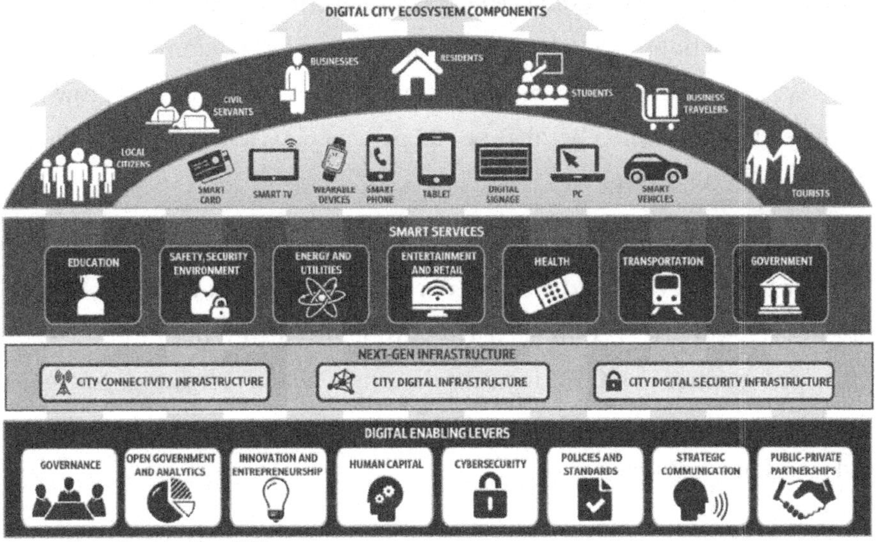

Source: Booz Allen Hamilton 2015

Smart Six: Infra, Buildings, Homes, Safety, Energy, Mobility

- **Smart infrastructure:** Smart infrastructure investments can transform traditional infrastructure. The global smart city ICT infrastructure market is projected to reach an estimated US$1.7 trillion by 2025, 17.5 percent CAGR versus 2017 (source: Verified Market Intelligence).

- **Buildings:** We spend 90 percent of our time in buildings, which account for around 40 percent of energy, 25 percent of water, and 40 percent of resource use, and a third of GHG emissions. However, just 10 percent of US buildings have basic smart technologies (e.g., efficient lighting or monitoring). The global smart building market is expected to grow from US$61 billion in 2019 to an estimated US$106 billion by 2024 (source: Markets and Markets, 11.7 percent CAGR).

• **Smart homes:** Smart home devices can control, automate, and optimize household functions like lighting, heating, cooling, entertainment, safety, and voice assistants. The global market for smart homes is projected to grow to an estimated US$405 billion by 2030, at a 25 percent CAGR from US$14 billion in 2015 (source: AT Kearney).

• **Smart safety and security:** Smart cities are integrating an increasing number of ICT-enabled devices and measures to improve safety and security. The safe city market will see an estimated US$226 billion in revenues in 2015–2021 with opportunities in CCTV, incident detection, adaptive lighting, emergency alerts, environmental monitoring, disease surveillance, epidemic monitoring, and smart/assisted care.

• **Smart energy:** Smart grids—digitalized and connected electrical grid comprising cleantech—could grow to an estimated US$169 billion by 2025 (source: Allied Market Research, 12.4 percent 2018–25 CAGR). New global investments in zero-carbon electricity generation capacity are expected to be US$3.8 trillion cumulative in 2021–2030 versus US$656 billion for fossil only (source: BNEF NEO 2019). Solar/wind power, smart meters, and electric vehicles could all be key catalysts for smart energy take-up.

• **Smart mobility:** The US$7 trillion mobility market is inefficient, expensive, and highly polluting (>20 percent of global emissions). A shift toward mobility as a service is a viable option to city car usage while digital capabilities (sensor data, diagnostics, analytics) can help improve efficiency and reduce costs (accidents, repairs),

Investments: Most Impactful Smart Cities Technologies

• **5G networks** to transform connectivity: 5G networks offer much higher data speeds, bandwidth, and reduced latency versus current 4G networks. 5G mobile handsets are arriving despite networks not being ready yet. Once 5G is properly deployed in four to five years, new services such as autonomous vehicles, IoT, and delivery drones can become feasible. There could be major transitions in some services (e.g., broadband + TV on cable) due to other transformations such as streaming TV/gaming explosion.

• **Urban transport transformation**—electric, shared, public: The biggest shifts in urban transport and mobility will be in the next decade. Electric cars to gain share versus gasoline cars in the 2020s. Ride-hailing to be much more disruptive due to electric vehicles (lower running costs) and regulation (city fossil bans, private car restrictions, emission taxes). Micro mobility (cycles, scooters) and multimodal mobility (different transport modes for one journey) could mean a redesign of cities (cycle lanes, pedestrian zones, traffic lights, sensors, etc.). Major transitions for sectors such as autos, oil and gas, construction, semiconductors, and software.

• **Circular, efficient and clean economy:** Circular economy will be a key focus for cities—plastic regulation (e.g., charges, bans), biodegradable packaging, and waste reduction. Corporates will have to consider a circular economy—for example, electric vehicle manufacturers will have to focus on battery recycling too. Sustainable packaging could be a big theme in the 2020s, as will be local production (e.g., vertical farming, speed factories). Sensors and analytics to boost energy efficiency and many major cities want high renewable energy share; for example, New York drove US offshore wind.

• **Internet of Things (IoT)** boosts data economy: Sensor and IoT deployment are mostly in cities, opening major opportunities for semiconductor, sensor, and software suppliers. For example, the expansion of the London Ultra Low Emission Zone for polluting vehicles will require cameras/sensors and software services. The smart cities' theme could become one of the biggest drivers of big data and AI growth in the 2020s. These new services would increase demand for cloud computing and AI.

• **Public spending surge:** Stagnating growth (esp. in the western world) and concerns about climate change impacts could lead to major fiscal spending to upgrade city infrastructure around the world. While monetary policy has been very supportive in the past decade, there has not yet been a major upgrade in spending and now the rising focus is on fiscal spending, especially in infrastructure, which could be a secular trend in the 2020s. And it is likely such spending will be directed toward smarter technologies, which can transform the urban landscape, for example, much more public transport, futuristic concepts like Hyperloop, and net-zero emissions buildings.

CHAPTER 11

Understand That People Are Not Numbers

Investors always think in terms of numbers. They track return on investment and investment risk. But entrepreneurs aren't thinking about that. They're focused on growing the business, growing relationships, reaching out to consumers, and making the product or service the best thing in the market. That's the ultimate goal, no matter how you look at it. When I say people are not numbers, I'm talking about human capital.

We've lost the dignity of human capital in our markets.

For example, our banking system is geared toward lending money. They can potentially give you a loan if you buy a property. But the if you default on that loan, they can begin the foreclosure process. The length of time you've paid them doesn't matter.

If you had a thirty-year mortgage and you've made all of your payments on time but in the twenty-fifth year, you just happen to lose your job and miss a payment, too bad. If there's a global pandemic or someone dies in your family, the bank doesn't care. They'll take your property because you missed one payment.

In any of those scenarios, human capital has lost its dignity because a person who is already down on their luck—they've lost their job, a family member, or whatever—and now they're going to lose the one property that they've had for twenty-five years. They're on the streets. Displaced. Homeless.

When an entrepreneur does not realize that our system is not perfect, then we've already gone down the wrong path. But when an investor or entrepreneur starts thinking that they will invest in people, then they are in the beginning stages of being an influencer. That's when you can have the biggest impact on people's lives. You can impact families for the greater good. That's something to take pride in.

Touching people's lives, influencing hearts and souls, should be the entrepreneur's number one priority. And that's what I mean when I say, "bet on people, not the numbers."

That's my personal approach.

The Influencers Who Influenced Me

My entire family has been an influence on me, my parents, and my grandparents. They did extremely well for themselves, and because of their success, I had a remarkable upbringing. I don't deny it, and I don't regret it either.

Being fortunate enough to grow up in that environment meant that I was also fortunate enough to not feel entitled to anything. I worked hard from a young age; I started at twelve years old with my grandfather and dad. I'd go to Dad's office in the summer and, essentially, be his intern.

You don't have to wait until you're in college to be an intern. Parents should start teaching their children to do things before they think they're ready. What the younger generations do today is breed a sense of entitlement. But they also allow their children to waste time, play video games, and laze about doing nothing. They don't encourage being a part of a team, participating in cross-cultural activities, or joining a community. That lack of connection won't help them in the future when they get into the real world and need to connect with people.

My upbringing taught me many things. First, to be humble. Even if we own restaurants, hotels, and other types of businesses, be humble. When I tell people the businesses I've owned, they often take a step back and say, "Wow!" But I don't believe in the wow factor.

Instead, I shrug my shoulders. "Great!" I say. "What are you doing about it?"

Before I learned to coach people, I learned to serve them. Before I learned how to delegate, I learned how to listen. And I started at a very young age.

When I say, "be a servant," I don't mean the lowly position of servanthood. I'm talking about providing a service. Being somebody who is a guide, or someone who provides advice. That's serving.

We should be proud of the process that turns us into leaders. My father and grandfather were both the biggest influencers in my life because, to me, they were huge figures. Successful people who took the time to instill in me the virtue and the value of service, and I had siblings, too.

Because I was one of the youngest among five siblings, I got to see how they managed their lives. I got to see how they managed their families and, quite honestly, that was a great influence on me. It nurtured my desire to become a better father, a better businessman, and a better husband. It wasn't easy, but I gave it my best. Since I had those influences and the desire to learn, I was able to parlay that into success in business, in my family, and in relationships in general.

Prosperous Companies Can Be Generous with the Community

When corporations do well, they tend to be generous. That's the human factor.

If the entrepreneur is successful, if their business has done well, and they take a friend to dinner, let's say the friend is unemployed, take him or her out and pay for the meal. Tell them, "I know you're having a tough go of it. Let me have the check."

That's the same DNA that corporations have when they are successful. Successful companies can be more philanthropic than companies that are not doing so well because they can afford it. They have to pay salaries before they can think about giving money away or supporting a charity. That's why I say successful companies can be more generous than broken companies, and it's true.

Every corporation should have a sense of social responsibility. It should be a part of the DNA, whether it's through the human resources department if they have one, or, if they're too small to have one, through the founders.

Social responsibility relates to helping communities with things. It can be public amenities or a service. Organizations may support mentoring children in impoverished communities while trying to teach those children that they can have a better future. It could be teaching them to stay away from drugs or alcohol. Or maybe they want to support social programs that help families stay together rather than separate because working single moms tend to have a harder time focusing on company goals. They're busy attending to the needs of their children and focused on concerns outside of the company.

Corporations must be aware of their social responsibilities. Entrepreneurial founders should think about how their companies will impact other people's lives. You have a life, your employees have lives of their own, and your customers also have lives of their own. It's important to connect to the human side of the equation. That will make employees more loyal to your company and more supportive of your company's goals.

Philanthropic work, charity work, foundation work, or any type of work that sustains a positive social environment and improves a community is the entrepreneur's social responsibility. And corporations also have that responsibility.

Public-Private Partnership and International Trade Agreements

Big government is a new social contract. Did you know that governments have committed US$8 trillion in the fight against COVID-19, which is more than the GDPs of India, Brazil, and the UK combined?

Key takeaway: growing surveillance, inequality, and the current inadequacy of health-care systems highlighted by the current crisis will act as a catalyst for change in politics, furthering populism trends and increasing the risk of social unrest.

Governments will exert greater influence on businesses across sectors, calling into question stakeholder importance, the role of privacy, and how we reward labor. Also, the boom in personal safety and pandemic preparedness funded by the government will continue beyond the immediate crisis above pre-crisis levels.

A private company wants to remain private for as long as it can. They don't need some government entity telling them how to run their operations, how to handle their finances, or where to invest their money.

Imagine that your corporate bank account is with Bank of America, and someone calls you up one day to tell you that your money should not just sit there. You must invest in blue-chip stocks or high-tech bonds. You'd likely respond, "No, I don't think so."

That's your money and you should be able to do whatever you want with it.

When the government tells companies what they can do, that's what we call "nationalization." It's the ugliest thing that can happen in a capital market. It happens in many countries around the world. Venezuela is the closest country, geographically, to the US where this happens. The naturalization process took over the whole private market and any public market access that was there before just disappeared. The stock market disappeared.

In the US now, we have PPPs, public-private partnerships. But we need to be careful how they work.

PPPs allow companies to invest in infrastructure, which includes schools, roads, and hospitals, or anything that a municipality might build. The entrepreneur happens to live in town and owns a corporation in that location. To support the community, they should invest in the infrastructure—roads, streetlights, potable water, electricity efficiency, air quality, schools, and so on. These amenities should be a part of the public-private partnership programs in every community to incentivize them to live better and improve citizen hygiene. It's to improve quality of life and health.

This trend is about participation or co-investment. The government invests a certain amount of money into those public amenities and the corporation invests a certain amount of money, as well. This is better than the government telling the corporation to give the government entity money so they do whatever they want with it.

We want to attract corporations to bring their businesses to the US and spend their money here. Let them invest in what they want.

The idea is, if you incentivize them to spend their money in the US, then they'll promote US products and support US initiatives. But they'll do that through, or in partnership with, the government entity.

Let's say I am a local manufacturer in the state of New York. New York is my primary market, but I eventually expanded to neighboring states because it makes sense for my business. Then I'll have a presence in Pennsylvania, Connecticut, New Jersey, and Massachusetts. Those would be the most likely states to expand my business into before going nationwide. When we get to that point, we can promote our products, made in the USA, to other countries. Then we'll be an international company.

Being an international company allows the corporation to grow by reputation while also helping your own country become more important in your sector.

Soriano Motori, for example, is an American company. We hire American workers. But we also produce electric vehicles in Italy. That makes us an international corporation that hires across the globe. And we impact people's lives in all the sectors in which we operate. So we pay taxes in the US and also in Italy.

Wherever we make money, we pay taxes. That's how it should be. If I compete with local manufacturers in other countries, there should be an equal balance of goods to each community we operate in. Goods that can be sold in the market.

Today, we see commercial deficits, what we call "trade wars," because Asian companies—companies making products in China, Korea, and Japan—sell in much larger quantities per capita to the US market. We have a population of four hundred million people. The Asian population is one billion. But Asian companies are selling three times the amount of goods to the American people than American manufacturers are selling to Asia. That gives us a clear disadvantage with a smaller population consuming more products from abroad than we are exporting to their larger population.

This imbalance should be corrected in what we call a "free trade zone agreement." If I manufacture out of Italy, I should be able to sell the same amount of product that is being sold in Italy by foreign companies. And vice versa.

COVID-19 Accelerates Digital Transformation for Work, Home, and Play

People around the world have been instructed to stay at home and work there where possible. This could have meaningful long-term impacts on companies that provide the technologies that enable this. We consider these in particular, in relation to remote working, the "stay-at-home" economy, and the rising penetration of e-commerce linked to this.

1. Working from home
2. The e-commerce and delivery economy
3. The "at-home experience economy."

Similar to the rise of new business models such as the sharing economy after the last financial crisis, and a wave of lasting innovation after the dotcom era, the post-COVID era will be defined by a combination of new technologies and companies enabling better services to eat, work, relax remotely (at home in particular in the short term), and the shift from offline to online being brought forward across all sectors and demographics. Thus, COVID-19 may accelerate the need for digital transformation.

Working from home: short-term shock or long-term reset of how we work? Perhaps the biggest test of technology has been the need to work from home where feasible, with COVID-19 causing most companies to temporarily close their offices. This had an immediate impact on companies facilitating working from home: video conferencing app Zoom saw average daily users rise from ten million to three hundred million from January to April 2020, for example.

While the number of people claiming to work mainly from home before the crisis was relatively low (e.g., 5 percent in the UK in 2019 [ONS]), it has highlighted which industries can adapt to it—and there are more than many previously expected. As our technology team notes in its Telecommute101 primer: in the US alone, the Bureau of Labor Statistics estimates that 60 percent of the 118 million salary-paid workers could work from home, in addition to the 5–7 percent that already has the capability.

The pandemic has demonstrated the potential for a distributed work revolution in the longer term, as companies and staff accelerate investments in working from home technologies and change working patterns to enable this transition. While it may not be applicable for every company or individual depending on circumstances, there is likely to be higher adoption of distributed/flexible working post-crisis, in our view, owing to a combination of the flexibility, commuting savings, and productivity improvements it offers employees, and the corporate overhead cost savings it offers employers. Beyond just travel to work, these technologies could also become instrumental in limiting or avoiding some business travel for certain sectors.

Exercise: Stay In, Work Out

Exercising at home was promoted during the pandemic to maintain physical and mental fitness; this gave rise to several digitally enabled alternatives to allow people to work out at home. Several of these could retain customers and cross-sell new services post-crisis.

Telehealth: the rush to get (doctors) online: With doctors and medical services largely unavailable during the pandemic to patients with non-urgent or life-threatening illnesses, virtual consultations rose significantly. UK-based Push Doctor recorded a 70 percent rise in consultations in the week the country entered lockdown, for example. More importantly, it saw rapid adoption from surgeries that were not previously connected to such platforms. This will have lasting implications for the supply of such services post-crisis, as a higher user base has downloaded or been made aware of the services through their health-care providers during the pandemic.

Moon shots accelerated: from quantum computing to Earth observation? Faster processing of more relevant data using supercomputers and AI could be accelerated with moon-shot technologies for commercial gain, including fighting future pandemics.

Quantum computing resources are being made available to researchers of COVID-19 free of charge by D-Wave, for example, and could become a crucial part of future detection of pandemics, accelerating commercial deployment more broadly.

Nanosatellites and space technologies could also be deployed to improve earth observation and communications (the UK/European space agencies have launched funds to investigate high-tech solutions in response to the crisis, for example). This increasing demand for data and bandwidth may also accelerate the need for new communication infrastructure such as 6G, with 5G bandwidth being at capacity sooner than expected.

Future food: crisis highlights the benefits of urban and vertical farming. A combination of lockdowns, panic buying of food, and restrictions on travel/imports has highlighted the challenges and global nature of the food supply chain and how it can be disrupted. This could accelerate more local produce, in particular, leveraging new technologies in vertical farming.

Singapore, in particular, which imports 90 percent of its food, reiterated its goal to produce 30 percent of its food needs locally by 2030, requiring a scaling up of urban farming as part of the solution. The crisis could increase the awareness and need for local produce, which could be more efficient and reliable with controlled vertical farming investments.

Paradigm Ten: Space

We entered the 2020s on the cusp of a new space age as commercialization by private companies enables humanity's liftoff from Earth. While only 560 plus people have ever been to space versus a seven billion-plus global population, we believe the 2020s will be the decade of space democratization with new technologies. Government spending on space is expected to reach $79 billion by 2025 versus $62 billion in 2016 and the space market opportunity involving nanosatellites, space tourism, and so on could reach nearly $1 trillion by 2030 (source: OECD, ULA).

Nanosatellite Launches Are Taking Off

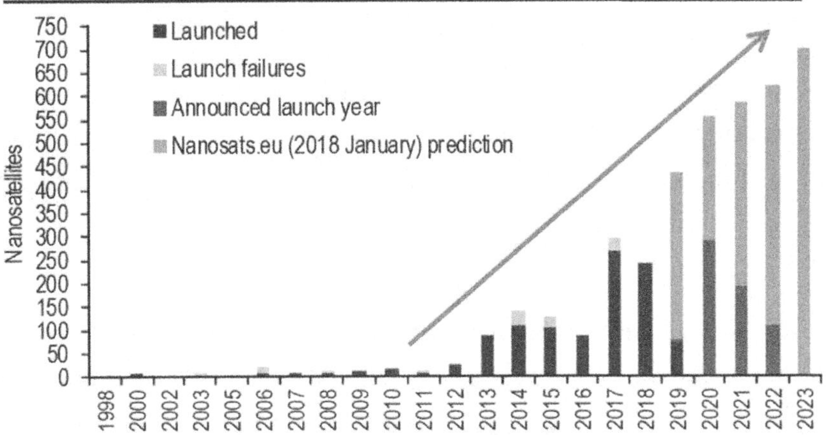

Source: Erik Kulu, Nonosats Database 2019

A radical new space age in the 2020s: In the next decade, we enter a completely new and radical space age that could far exceed the 1960s missions to the moon. New technology (reusable rockets), innovative speedy private companies, miniaturization of electronics, and new services (Internet from space, space tourism) could revolutionize space like never before. In the 2020s, large privately built rockets and shuttles could send humans back to the moon (only 24 people have visited the moon, with the last one flying in 1972) and even put humans on Mars (Space X's ambitious Mars targets: cargo 2022, cargo and crew 2024).

Longer-term, new space technologies could boost innovation, just like the Moon missions did for computing.

Near term, nanosatellites could deliver worldwide broadband to nearly 40 percent of humanity not yet connected to the Internet as well as accelerate 5G networks and new services. With more countries and corporations getting involved in space/satellites due to lower costs and more functionality, the next decade is set to be the most exciting ever for space.

• **Space Age 2.0: privatization of space:** Historically driven by governments, many private enterprises have started providing space services to governments and developing radical new technologies. Private companies benefit from focused leaders, less bureaucracy, commercial acumen, and blank slate design.

• **Billionaires' playground:** This new era of private space technology development is being driven by billionaires, most notably Elon Musk and Jeff Bezos. Sixteen of the world's five hundred billionaires have space investments, which could grow from a US$360 billion market today to potentially as big as an estimated US$2.7 trillion by 2045 (source: ULA).

• **Costs fall due to new technologies:** Due to reusable rockets, the cost of space launches could fall by a factor of one hundred (SpaceX, Blue Origin, Virgin Galactic, etc.). Much bigger rockets, such as Falcon Heavy and Starship (in the future), can cut costs further. Intercity travel using rocket technology could also become effective.

• **Wider support:** Space Age 2.0 also benefits from the involvement of many new countries (80+ countries with satellites in orbit), falling launch costs (Rocket Lab US$5 million), rising start-up investments (US$16 billion since 2000, US$2.8 billion in 2016) and a more favorable policy backdrop (various countries pushing for space tech).

Space Age 2.0 in a Nutshell

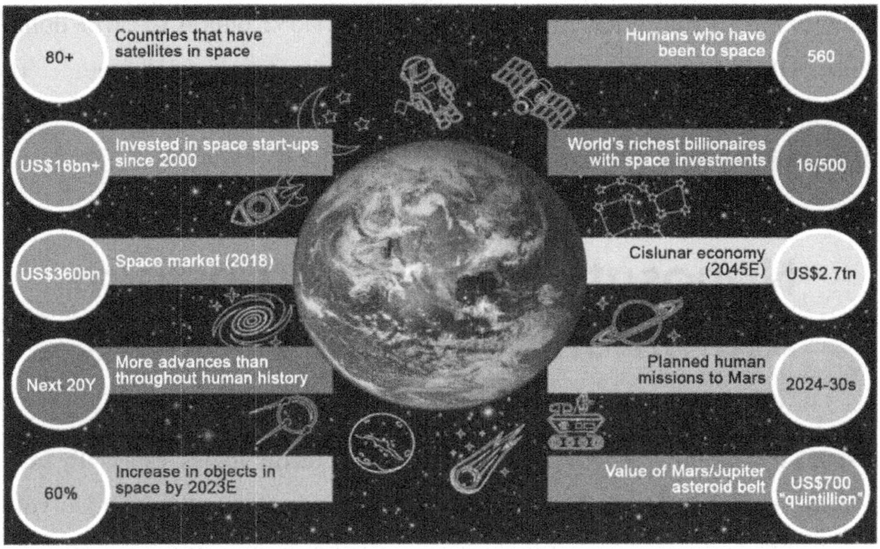

Source: BofAML Global Research based on various sources

Reusable Rockets – Technological Leap, Huge Cost Reduction

• **Reusable rockets game-changer:** Reusable rockets reduce rocket launch costs by an order of magnitude. Onetime use rockets were too expensive for wider use as the first stages of rockets are expensive relative to the fuel costs. Recovery and reuse of rockets can significantly reduce costs per launch.

• **Private companies delivering innovation:** SpaceX landed a reusable rocket on land in December 2015 and then on a drone ship in April 2016. SpaceX first reused a recovered rocket and landed it back in March 2017. The space company Blue Origin was the first to launch and recover a rocket in November 2015.

• **Costs down dramatically and to fall further:** The Saturn V, which took humankind to the moon in 1969, cost almost US$2 billion in 2018 dollars versus US$95 million for SpaceX's Falcon Heavy launch in 2018 (source: The Guardian). This implies a less than 90 percent cost reduction in real terms since the moon missions, with more savings to come due to reusability.

Nanosatellites Can Transform Communications & Business

• **Space has already been transformative:** Space technologies, especially satellites, have already greatly benefited humanity in communications (video/voice calls), entertainment (TV), inflight Wi-Fi, weather monitoring, crop monitoring, climate change research, and mapping (satnav/GPS), among many other use cases.

• **Nanosatellites could win by sheer numbers:** NextGen nanosatellites could be transformative due to their sheer numbers. Nanosatellites can weigh only a few pounds, helped by the miniaturization of electronics. Small size and falling launch costs can allow thousands of nanosatellites to be launched into low orbit. Planned launches of nanosatellites would dwarf the current five thousand or so satellites in orbit (targets: SpaceX's Starlink 12,000 nanosatellites, OneWeb 600–700, and Amazon >3,000). Nanosatellite webs can link the whole world, enabling fast connectivity even in remote areas.

• **Potential for disruption**: Nanosatellite constellations may enhance current networks but also disrupt them. Worldwide satellite networks could bring many more people into the connected world and unlock many services like mapping, surveillance, autonomous cars, and yet unknown services.

Space Tourism for Rich But Intercity Rocket Travel Promised

• **Space tourism for the rich:** Taking tourists into space is already close to reality. SpaceX has already agreed to take Japanese billionaire Yusaku Maezawa on the first space tourist trip to the moon, with the mission expected in 2023 (Mr. Maezawa plans to take 6–8 artists with him). Seven people have paid to go to space before (e.g., to the International Space Station) but no space tourist has flown since 2009. This is seemingly about to change with several companies, including SpaceX, Blue Origin, and Virgin Galactic, close to offering space tourism, having already successfully demonstrated their rocket capabilities over multiple launches and recovery. Space tourism, however, will be confined to the rich until costs fall.

• **Intercity space flights could be more transformative:** Launch of commercial intercity travel via space could open up major opportunities, dependent on cost reduction. If ten- to twelve-hour long-haul flying times between global cities could be cut to two to three hours, there could be a market for such services, especially at the high-end bracket. If costs fall enough to compete with current (profitable) business travel, there could be disruption to the airline industry.

Investment: Space Technologies Most Impactful for the 2020s and beyond

The first phase of nanosatellite network deployment is expected before 2025, which would not only help with the learning curve but also allow new services to be launched, some of which could become disruptive and widely embraced.

Nanosatellites could also lead to significant increases in surveillance capabilities and therefore inhibit privacy further, becoming a political issue. Major improvements in intercity space travel could open up the long-haul business trip market for disruption and influence globalization.

CHAPTER 12

Impact Investment Versus Social Responsibility

Our societal welfare, economic or otherwise, is built on having a healthy and thus productive population. The pivotal significance of health care and pharmaceutical products for society makes the industry highly important to stakeholder expectations, as evidenced by COVID-19.

For stakeholders, a comprehensive ESG framework not only calls for strong environmental initiatives but also businesses that embrace its wider social impact, diversity, and inclusion.

Nevertheless, the pharmaceutical industry in recent times has been plagued by scandals that have eroded public trust, for example, governance, product safety, opioids misuse, and drug pricing. COVID-19 gives the sector a shot at redemption as it becomes the key pillar to prevent future pandemics and will likely become one of the most important ESG sectors in the coming years.

Social: The "S" Pillar of ESG Will Gain Traction

We believe the consequences of the coronavirus will result in growing investor focus on companies' social factors. We expect investors to consider more qualitative indicators as a gauge of long-term competitiveness (human capital) and operational integrity (business continuity). These indicators include, among others:

- Employee health benefits (e.g., paid sick leave, health and well-being programs, short-time work, telemedicine)
- Emergency response/business continuity (e.g., telecommuting, distributed management, backup working office sites)
- Supply chain controls (e.g., quality and safety management, local vs. global dependence)

Why does the "S" matter? Companies allocating resources to the COVID-19 crisis are likely to foster employee and community goodwill to enhance brand and reputation. Companies with a record of good employer-employee relations have markedly outperformed peers during the coronavirus crisis.

Overall, there's been a sea change—from corporates focusing solely on shareholders to a broader focus on stakeholders, including employees, customers, suppliers, and communities. Companies around the world have stepped up their efforts to fight the virus and demonstrated how the private sector is augmenting the public sector response to COVID-19.

Surveys by several analysts flag that the "S" of ESG has been the number one most influential factor for their companies during the crisis, underpinning the social importance of stronger health-care systems for the future. Two-thirds of our analysts highlight that their companies will see higher costs on employee benefits, for example, health care, hygiene, safety, insurance after coronavirus. Half believe that public health systems will see a larger allocation in national budgets, with 40 percent believing there will be more data sharing with government agencies.

COVID-19 is Forcing Political Trigger Points

The COVID-19 pandemic has brought various trigger points to the forefront, increasing the risk of social unrest and populist policies:

- Surveillance: the option to monitor citizens through phones and other connected devices has openly been combined with governments' abilities to control and limit behavior. The differing approaches in various countries over a government's role in data collection and who should have access as liberties and individuals' rights to privacy are encroached upon.
- Inequality: is set to grow from COVID-19. Most highly paid individuals have been able to work from home, while a higher proportion of lower-paid workers have either lost their jobs or work in dangerous, exposed conditions, making COVID-19 the great polarizer. The negative impacts for the most disadvantaged will be felt for years to come as education, health care, and work access, among others, are limited for those demographics.
- Health care: the inadequacy of health-care systems and lack of preparation by governments for long-term risks have been exposed in this pandemic. COVID-19 will amplify health care's social role in providing a healthy workforce and accelerate other pressing global public health issues, such as drug pricing, antibiotics resistance, pandemic prevention, and funding of health care.

Combined, these three flash points pose a threat to current political positions, potentially forcing more social unrest and creating pressure for populist policies. We expect governments to 1) continue to use technology and surveillance, 2) come under more pressure to help workers, 3) exert more influence on companies, switching the focus from shareholders to stakeholders, and 4) improve their preparation for risks, at least for a few years.

Many investors, when they first hear about impact investing, wonder if it's the same as socially responsible investing. No, it is not.

Socially responsible investing applies a set of positive and negative screens to a group of publicly traded company stocks. For example, a mutual fund may avoid investments in tobacco, alcohol, or firearms. Impact investing goes beyond these passive screens by actively seeking out companies or projects with the potential to create positive economic, social, and/or environmental outcomes. That's the difference between socially responsible investing and impact investing.

Fund managers of socially responsible investments focus on picking assets that do no harm, whereas fund managers of impact investments seek to make a positive difference in the assets they choose.

In seeking to make a positive impact, impact investing fund managers measure and report their impact in a transparent way. Going beyond that, there are also environmental, social, and governance issues (ESG) that are relevant to company operations and business growth strategies. ESG analysis considers the effect that a company's operations have on the market, a particular community, or the environment. It's becoming more mainstream and, in recent years, ESG analysts have focused on long-term value that reduces brand and reputational risk.

The United Nations has something called Principles of Responsible Investment (PRI), which provides details on the best practices for ESG investing. The last time I read it, there were signals for more than seven hundred investment managers and asset owners globally. Corporations on the list included BlackRock, KKR, CalPERS, and many more. Apple and Alphabet have joined forces to help track individuals. The companies aim to use Bluetooth to allow interoperability between various health apps across countries to make tracking more consistent, before coming out with their own contact tracing platform.

Signatories to the PRI include impact investment stakeholders, institutional investors, and investment managers such as the Rockefeller Brothers Fund and BNP Paribas Capital Partners. They wanted to create a set of common social and impact metrics to increase transparency and the credibility of impact reporting, which led to the creation of the Impact Reporting and Investment Standards (IRIS) framework.

IRIS is a metric that catalogs the taxonomy of the impact assessment framework. Each borrower must select at least one impact from a list of nineteen objectives provided. UN standards, called global sustainable development goals, list seventeen objectives that were approved by 194 sovereign states. These are reviewed every year. Regarding impact measurement, there's due diligence and monitoring. These are done the same way as with ESG. Due diligence includes adherence to the IFC Exclusion List, position statements, and a commitment to a reputation for sustainable and ethical business policies and practices. The core metrics for impact include job creation, wages, growth, increase in revenues, net profit increases, and increased taxes for local governments.

The IRIS framework applies across sectors and geographies to give investors a standard measure of nonfinancial impact on their investments. This typically ranges from average employee wages to metric tons of greenhouse gases. That's why you have large corporations, like McDonald's, paying a higher base salary than their competitors.

Areas Where Impact Investing Makes a Big Difference

There are a growing number of investors who want to see their money go toward stocks or funds of assets that are both profitable and reflective of their social values. I am proud to say that I am one of them.

The Great Polarizer: Gap Grows Between Haves & Have-Nots

COVID-19 has further exposed inequalities in society, pressuring not only governments but companies in terms of their behavior and responsibility to their employees. Despite claims by many that COVID-19 has been a great leveler, with anyone susceptible and all feeling the costs of lockdown, the reality is very different.

Not only are older demographics and those with underlying health conditions more affected, but the way we are sharing the burden of protection is also unequal. Those who can work from home are typically higher earners, while manual labor and lower wage earners are more likely to be either losing jobs or having to work in dangerous, exposed conditions.

The US has already experienced protests about lockdown, and the pressures exerted on parts of society are likely to increase the risk of social unrest.

Some of the objectives associated with impact investing include better access to education, which includes business activities that actively seek to provide schooling to students who have never been in school. Soriano Motori is going to have its first international intern this summer. She will travel between Los Angeles, Miami, and New York City as well as our offices in Italy and Monte Carlo, France.

This young intern comes from Michigan. She went through a series of interviews with management, and everybody liked her, so we hired her. She started on June 15, 2021. She'll be with us for twelve weeks and it will be the ride of her life.

Sometime soon, I'm going to set up a foundation where we raise capital to offer full scholarships. We'll invest in specific sectors in the market, and we hope recipients of these scholarships will become future employees of ours. In a way, you could say we nurture education through those scholarships, and we're nurturing future employee loyalty to our companies.

Another objective is access to energy. Through our business activities, we seek to provide electricity as an alternative source of energy for vehicles. That's a part of the approach to our electric vehicle platform. We believe electricity can be the answer to reducing greenhouse gas emissions.

In Los Angeles, when social distancing guidelines and business shutdowns led to decreased traffic on the roadways, that was the first time many people could actually see a clear sky. They began to appreciate how beautiful their city is, but the smog typically obscures it.

Soriano Motori aims to help with regard to environmental impact. We're a small company and it's not necessarily a big move compared to the global output of emissions, but it's something that will help. Someone once told me a story about a man who took a walk on the beach with his son, and they discovered thousands of starfish on the seashore. They'd been washed up on the shore by the tide.

We know that starfish die outside of water. The man's son grabbed the starfish, one at a time, and threw them back in the water, but there were thousands of them. After he'd thrown a few, his father asked, "What are you doing, son? Throwing them back in the water won't make a difference."

The son replied, "It will make a difference to the ones that end up in the water."

That's my philosophy when it comes to environmental impact. Regardless of size, I believe Soriano Motori can become influencers to larger corporations just by doing the right thing.

Around 6.7 percent of global working hours are set to be lost in the second quarter of 2020, the equivalent of 195 million full-time jobs. Moreover, 37.5 percent of global workers are in the four most hurt sectors—food and accommodation, retail and wholesale, business services and administration, and manufacturing (source: ILO 2020).

Another impact objective is access to financial services. We seek to provide individuals and organizations access to financial services that, before, have not been available to them. In our company, when employees want to buy a home or expand educational opportunities for their children, we can help them do that.

Affordable housing, for instance, and access to new markets.

We want to foster activities that enable access to new markets for products and services produced and sold by the company. If you work for an electric vehicle company, for instance, I'd expect you to also drive electric vehicles. Otherwise, why are you working there? Just for a paycheck?

With regard to access to new products, agricultural productivity is another area where substantial potential exists. Soriano Motori wants to increase the number of agricultural products and services produced and sold by the company. Again, we struck a deal that positions us as the only family in Australia to distribute olive oil throughout Asia. That's something we've been cooking up for almost twenty years and which has finally come to fruition.

Capacity building. We want to train and provide technical assistance to individuals within our organization.

Community development is another impact objective important to us. We want to provide financially profitable products to end-users in our local communities.

We also want to promote and increase employee ownership of the corporation. We have incentives that offer equity plans in the company so that employees can be part owners.

Energy conservation is also important to us. We seek to invest in techniques that reduce the amount of energy needed to carry out current processes and tasks, environmental conservation, equality, and empowerment. That's a big one for us.

We want to promote equal access to employment opportunities within the organization, products, and we want to be on the cutting edge of food security. That is, we want to increase the number of individuals, or households, who have access to sufficient food to maintain a healthy lifestyle because a healthy employee will always work better. That feeds into health and wellness. Anything that sustains and improves a healthy lifestyle.

Job creation. We seek to increase the total number of full-time and part-time employees within the organization. We hope to move them from being outside consultants to positions within the company.

Productivity and competitiveness are other big ones. We want to increase the number of products and services we produce.

Pollution prevention and waste management are also key elements. We're very interested in recycling. That means the collection, transport, treatment, and disposal of waste. It also means the control, monitoring, and regulation of the production, collection, transport, treatment, and disposal of that waste. Thirdly, it means the prevention of waste production through in-process modifications, reuse of materials, and recycling. It starts with paper. It also includes ink for our computers. Large corporations don't care about that anymore, but as a small company, we do.

We also want to increase the value of employee wages, including bonuses. We don't exclude part-time employees from benefits. We're pretty good at including everyone.

We don't have fifty thousand employees. We have five hundred in all of our silos around the world. That's five hundred families that I am responsible for and wish to make an impact on. If I do that well, the impact will ripple into the communities where they live and into the rest of the world.

These are the impact objectives Soriano Motori has internally and I think most entrepreneurs can execute the same or similar objectives without much cost.

Social inequality issues growing in importance: Highly valued organizations have faced a backlash for their behavior in this crisis; Harvard University recently returned $8.7 million in federal aid after political pressure, while Disney has faced negative media attention for maintaining its executive bonus schemes while cutting pay for one hundred thousand workers.

Three Styles of Investing

I mentioned these earlier, but there are three metrics to impact investing every entrepreneur should know. These are environmental, social, and governance (ESG); socially responsible investing (SRI); and impact investing.

Environmental, Social, and Governance

The ESG framework is based on convention. There's a material effect on the long-term risk and ROI for each investor. The investment fund manager integrates ESG analysis into their due diligence. That filters into asset class decision-making, portfolio management, and reporting. That translates into training for investment partners. The fund manager gathers relevant ESG data, which enables investors' evaluation of policies and procedures and ends up impacting borrowers and equity holders in the companies they invest into.

ESG due diligence and monitoring is a four-step process. It starts with screening and moves into due diligence, monitoring, and reporting.

Screening is adherence to the IFC exclusion list and the industry position statement. It also involves a commitment to sustainable and ethical business policies and practices.

Due diligence looks at the geographical and industry context, compliance with local legal and regulatory requirements, adherence to any relevant local or international voluntary or certification standards, compliance with international ESG best practices—specifically, environmental and social management systems (ESMS)—and the IFCs, environmental, and social performance banners.

Monitoring relates to activities that go against the IFC Exclusion List, compliance with local legal and regulatory requirements, adherence to relevant local and international voluntary and certification standards, and so on.

Periodic and annual reports on portfolios must be filed that also address the borrower company ESG policies, practices, and performance highlights.

Governments just got bigger as pressure to step up grows. Governments are adopting a "hands-on" approach: fiscal stimulus, greater regulation, and more surveillance mean they have increased control. Higher deficits, social spending, and fiscal stimulus could become the government mantra of the next decade. Countries have resoundingly chosen to help and protect communities, including the most vulnerable in society.

Consequently, schools of thought that have been the subject of debate and interest over the past decade are, in effect, being implemented: modern monetary theory, radical money printing, the return of Keynesian economics, and fiscal over monetary stimulus.

Provided we don't see runaway inflation as a result of these policies, countries may continue to use these economic mechanisms to ease the pain of the ensuing recession. Longer term, a willingness to spend to prevent pain for workers could mean greater openness to the concept of universal basic income and the expansion of welfare states.

Businesses are likely to see further scrutiny to act more in the interest of all stakeholders (employees, society, customers, etc.) rather than a sole focus on shareholder returns as governments provide aid and back loans.

Impact Investing

Impact investing is about improving society; that's why the government gives away free money. It's about helping businesses or organizations complete a project, develop a program, or do something positive that benefits society.

If you had an alpaca on your property, for instance, or any type of animal within a certain amount of square feet, then you could consider some kind of sustainable farm. Every country has a different framework and the entrepreneur would have to engage in research to determine what category they fall into.

This crisis has resulted in an acknowledgment that a company's responsibilities and purpose go beyond providing returns to the shareholder. This is not a new trend. The growing ESG focus of asset managers and the 181 signatories, including BlackRock and JP Morgan, of Business Roundtable, that "share a fundamental commitment to all our stakeholders" is evidence that this trend is also entering the financial industry (source: Bloomberg). The current crisis has shown material examples of that change in priority to other stakeholders:

- The UK's largest banks, HSBC, RBS, Barclays, and Lloyds, have stopped dividends and buybacks after financial regulators pushed them to preserve more capital given the "unique role that banks need to play in supporting the wider economy" (source: Bank of England Prudential Regulation Authority 2020). Similarly, the AFEP (French Association of Private Enterprises) has instructed its members, such as Michelin, to cut dividends when taking any form of state support.
- Industry has been requested to change its factories to increase production of ventilators and other vital products. GM, for instance, was

compelled to produce ventilators extending the DPA (Defense Production Act) beyond its typical use for military procurement.

- Thirty-three percent of BofA analysts expect shareholder rights to be eroded either through government intervention or changing priorities within companies. A further 24 percent expect temporary change to the rights of shareholders in some sectors as governments earn influence over business in exchange for the aid handed out.

It's important for borrower companies to align with the mission of fund managers. That is, they must create a positive, measurable impact. Therefore, they must self-identify and provide baseline data for at least one company's specific impact objective.

Socially Responsible Investing

Did you know that nursing is one of the fastest-growing professions in the US, and one of the least likely to be automated by robots but it is one of the least well paid?

Key takeaway: Health care will likely be the most important ESG sector in the coming years. COVID-19 will amplify the importance of its social role and accelerate other pressing global public health issues, such as drug pricing, antibiotics resistance, pandemic prevention, universal vaccines, and so forth.

SRI involves actively removing investments or choosing investments based on specific ethical guidelines. It's as much about what the fund manager removes as it is what they choose.

You look at all the different impact objectives I mentioned then go through the process that I described earlier for ESG. That's the framework, and based on that, you decide what you're investing in.

For example, in the agricultural or agro-processing sector, we performed a study where we had financed six different businesses. There were more than 1,000 permanent jobs created. Of those, 340 were jobs for women and there were 607 employees trained with more than $31 million invested; that involved about fourteen transactions.

That's one example of how a breakdown of metrics would work. The typical investment is a term loan. You have a certain number of years to pay it off and then you invest X amount of dollars with an expected return over five or ten years.

The loan could be for anything. We financed a grain processor, for example. We also financed a frozen bakery products manufacturer, a waste-to-fuels processor, a power producer, and a ship maintenance and repair service provider. These are just a few examples of sectors where an investor may look and use the abovementioned parameters as metrics.

Remember, you're borrowing money. So, if the entrepreneur wants to make waves, the ESG framework is good because you can access capital through these metrics. And you can get a large amount of money for free.

One application is funding with government entities that support the ESG framework. That's pretty much everything we're investing in today.

On another note, the areas that impact these are:

- Big data and AI: Cloud, data storage, semiconductors, telecom equipment providers, and data analytics companies will be necessary to store, process, transport, and analyze the larger volumes of data resulting from the crisis.
- Surveillance: Companies that enable surveillance and data gathering, such as cameras, sensors, and Internet of Things producers, are likely to experience continued demand to expand networks as companies and governments increase dependence on them.
- Cybersecurity: As the world moves online at a greater pace, the risks relating to the online system also increase. Companies have seen some of the fastest expansion of their networks as working from home involves a much larger number of personal-employee devices and network systems. The pace of change and increase in business exposure to

the Internet, new devices, and larger networks will create more cyber vulnerabilities to be exploited but also protected.

- Life sciences: Lab equipment, diagnostics, food, and consumer product safety, among others, are clearly necessities in a world where viruses can spread easily. While a vaccine or herd immunity is not achieved, the risk of second, third, and further waves of COVID-19 outbreak means high volumes of testing will be needed long after the current peak has declined.
- Biotech and pharmaceutical sectors have previously been excluded by ESG investors, given their use of animal testing and other issues. However, their key roles in providing vaccines and other support may result in a change of attitude.
- Testing Inspection and Certification (TIC): COVID-19 has highlighted the inherent risks within our globalized supply chains and interconnected world. Similarly, to the increased regulation and transformation that the financial industry experienced post the 2008 financial crisis, companies will need to better prepare for risks and may need to prove better compliance relating to health and environmental factors. To begin preparing requires an understanding and assessment of risks and companies that provide testing, inspection, and certification services will be key enablers. Food and animal surveillance could be deemed more necessary—most diseases are transmitted from animals (source: Lancet). Further, hazardous medical waste is likely to receive greater scrutiny and care to prevent a relapse of the disease in the future—perhaps even years later.

Why I'm So Passionate about Helping Entrepreneurs

My passion is defined by my personality, my own spontaneity. I'm very spontaneous. I'm also impetuous, which has a negative connotation, but I've always been that way.

One example I'll give is from childhood. When I was eight years old, Europe had a bicycle competition for anyone at least twelve years old. My brother was twelve, but I wasn't. I was like his shadow. I remember that he signed up for this race. I also remember wanting to be there, and I wanted to go so badly even though I hadn't signed up for it. So I showed up with my little bicycle with my mother behind me saying "You can't do it, Marco, because you're too little."

When I got to the starting line beside my brother, my mother was yelling at me, "Marco, what are you doing?"

I said, "Mom, I'm competing."

The officials managing the race told my mom not to worry. "He'll be fine," they said. But Mom started crying. I just wanted to be a part of the race, to compete, and to be beside my brother.

When the race started, led by a truck out front, there was a guy with a megaphone yelling out the countdown, and a big *boom* got us going. I started pedaling like there was no tomorrow. It was a super small bike trying to keep up with these larger bikes. In the late 1980s, BMX racing bicycles were very popular. But I was on a little crap bike with tiny wheels and pedaling like it was the last day of my life.

In the middle of the race, my brother's chain fell off of his bicycle. I could see that he couldn't go any farther. About one hundred yards in front of him, I stopped. I hit the break, *The Fast and the Furious* style. I went back to my brother, got off my bike, and helped him put his chain back on. Then we got back on our bikes and finished the race.

We didn't win. Not even close. Crossing the finish line, we were somewhere around seventy-four and seventy-five, or maybe eighty-four and eighty-five. But a journalist saw what I did and stopped to interview me. Even though I didn't win, they lifted me in the air to celebrate my heroics. I had no idea what they were doing or why, but I was hoisted up on people's shoulders and carried around like a champion.

They carried me to the winner's circle and gave us medals. Both my brother and me. Kellogg's was the sponsor of the race, and they gave us a five-year supply of cereal. Of course, that made me happy.

Through that experience, my brother showed me how to be resilient and competitive. He taught me to be strong-willed and passionate about what I'm doing. It just happened to be on bicycles, and today I manufacture motorcycles.

I don't know what that story means other than it gave me the impetus I have today for helping entrepreneurs succeed. First, I have the drive to succeed on my own, to overcome obstacles, to meet challenges, and to compete until I win. I want to help others do the same. That's what this is all about, and I hope it inspires you.

Now, my last remarks for your wall:

Top 10 Results: Permanent Change From a Temporary Crisis

1. Revenue recovery to take years: Over half of analysts think it will take at least two to three-plus years for revenues to recover to pre-COVID levels. Banks, energy, capital goods, transportation, and aerospace and defense analysts overwhelmingly thought that revenues would take two to three years to recover, while health care, tech, consumer staples, and telcos expected zero to one year.
2. Capital structure changes for the long run: Only 15 percent of analysts expect no changes to their sector's capital structure. Instead, nearly

half expect deleveraging, and a third expect fewer buybacks and dividends over the next two to five years.

3. Supply chain change: 37 percent of analysts with relevant coverage have officially seen, or are expecting, changes to supply chains through either reshoring or diversification.
4. Beneficiaries and challenges: fundamental analysts expect the beneficiaries to be tech, health care, e-commerce; and the challenged to be air travel, hotels, and bricks and mortar retail.
5. Climate change still on top: Half of analysts still expect climate-friendly investments to increase post-COVID, while only a quarter expect them to decline.
6. Robots and automation: 43 percent of analysts with relevant coverage have already seen or expect changes to or acceleration of automation and robotics plans.
7. Top trends to continue post-COVID: A majority of fundamental analysts expect to keep using the following trends: offline to online, virtual work, food delivery, automation, and streaming.
8. Consumer caution: When asked about their own behavior, over half of fundamental analysts expected to be more cautious about using shared goods post-crisis, 36 percent expect online to become their default, and a third will travel less.
9. Preparing for the next disaster: Beyond building cash, companies are preparing for the future black swan or pandemic by building backup sites and diversifying supply chains.
10. The next black swan? Social unrest and COVID 2.0?

Appreciation from the Author

"There are only two ways to live your life: One is as though nothing is a miracle. The other is as if everything is. I believe in the latter." —Albert Einstein

We've reached the end of our journey in this book. I thank you for your kindness in investing time to internalize my words.

There is no such thing as a unique scientific vision, any more than there is a unique poetic vision for entrepreneurs. Starting the journey of your idea is a mosaic of partial and conflicting visions. But there is one common element in these visions. The common element is rebellion against the restrictions imposed by the locally prevailing culture and markets, as the case may be.

With that, please consider sharing your journey with me. I would welcome hearing your feedback and notions. You may reach me at 347.907.1214 or via e-mail: marco@sorianogroup.com.

Kindest regards,

Marco Antonio Soriano IV

Made in the USA
Monee, IL
05 November 2026

67816263R00118